CELL BIOLOGY

LABFAX

The LABFAX series

Series Editors:

B. D. HAMES Department of Biochemistry and Molecular Biology, University of Leeds, Leeds LS2 9JT, UK

D. RICKWOOD Department of Biology, University of Essex, Wivenhoe Park, Colchester CO4 3SQ, UK

MOLECULAR BIOLOGY LABFAX

CELL BIOLOGY LABFAX

CELL CULTURE LABFAX

Forthcoming title

IMMUNOCHEMISTRY LABFAX

EDITED BY

G.B. DEALTRY and D. RICKWOOD

Department of Biology, University of Essex,
Wivenhoe Park, Colchester CO4 3SQ, UK

ACADEMIC PRESS

First published in the United Kingdom 1992 by
BIOS Scientific Publishers Limited
St Thomas House, Becket Street, Oxford OX1 1SJ, UK

ISBN 0 12 207890 X

A CIP catalogue entry for this book is available from the British Library.

This Edition published jointly in the United States of America by Academic Press, Inc. and BIOS Scientific Publishers Limited.

Distributed in the United States, its territories and dependencies, and Canada exclusively by Academic Press, Inc., 1250 Sixth Avenue, San Diego, California 92101 pursuant to agreement with BIOS Scientific Publishers Limited, St Thomas House, Becket Street, Oxford OX1 1SJ, UK.

Typeset by Unicus Graphics Ltd, Horsham, UK.
Printed by Information Press Ltd, Oxford, UK.

PREFACE

Cell biology has been an important area of the life sciences since the invention of the microscope enabled biologists to look at the various structures and levels of organization in cells. However, following the enormous surge of interest in molecular biology, cell biology had tended to become overshadowed by research into molecular aspects of biology. Recently there has been a resurgence of interest in cell biology as molecular biologists look for cellular functions of gene products and with the realization that substantial areas of biology remain beyond the reach of definitive investigations at the molecular level.

The task set to the editors was to distil the essential data needed by people currently working in the main areas of cell biology into a single manageable book. This task was made somewhat easier following a decision that topics in plant cell biology should be covered in a separate book. Even so we were confronted with difficulties in selecting, compiling and balancing contributions that would be of interest to the majority of cell biologists on an almost daily basis while omitting dated and less relevant material. In this task we have been greatly helped by authors suggestions and the readiness of contributors to modify their texts to meet the needs of the book. We hope we have achieved a correct balance of the various facets of cell biology and that this book will prove useful to everybody working in the area of cell biology.

G. B. Dealtry

D. Rickwood

HAZARD WARNING

Some of the chemicals and procedures described in this book may be associated with biological hazards. In addition, the reader should be aware of the hazards associated with the handling of animal tissue samples. While efforts have been made to indicate the hazards associated with the different reagents and procedures covered in this book, it is the ultimate responsibility of the reader to ensure that safe working practises are used.

CONTENTS

5. DNA AND CHROMOSOMES OF CELLS 67

6. THE COMPOSITION AND STRUCTURE OF MEMBRANES 77

CONTRIBUTORS

P.D. BROWN

Department of Physiological Sciences, University of Manchester, Stopford Building, Oxford Road, Manchester M13 9PT, UK

G.B. DEALTRY

Department of Biology, University of Essex, Wivenhoe Park, Colchester CO4 3SQ, UK

A.C. ELLIOTT

Department of Physiological Sciences, University of Manchester, Stopford Building, Oxford Road, Manchester M13 9PT, UK

J. GRAHAM

MIC Medical Ltd, 131 Mount Pleasant, Liverpool L3 5TF, UK

N. HARRIS

Botany Department, University of Durham, South Road, Durham DH1 3LE, UK

R.E. LEAKE

Department of Biochemistry, University of Glasgow, Glasgow G12 8QQ, UK

R.P. NEWTON

Biochemistry Department, University College of Swansea, Singleton Park, Swansea SA2 8PP, UK

M.K. O'FARRELL

Department of Biology, University of Essex, Wivenhoe Park, Colchester CO4 3SQ, UK

D. PATEL

Department of Biology, University of Essex, Wivenhoe Park, Colchester CO4 3SQ, UK

D. RICKWOOD

Department of Biology, University of Essex, Wivenhoe Park, Colchester CO4 3SQ, UK

D. SCHULSTER

NIBSC, Blanche Lane, South Mimms, Potters Bar EN6 3QG, UK

ABBREVIATIONS

A-II	angiotensin II
Aces	*N*-(acetamido)-2-aminoethane-sulfonic acid
ACTH	adrenocorticotropic hormone (corticotropin)
ADA	*N*-(2-acetamido)iminodiacetic acid
ANP	atrial naturetic peptide
ATP	adenosine triphosphate
BDMA	benzyl dimethylamine
Bes	*N*,*N*′,-bis(2-hydroxyethyl)-2-amino-ethane-sulfonic acid
BSA	bovine serum albumin
cAMP	cyclic adenosine monophosphate
CBG	cortisol-binding globulin
CCD	charge-coupled device
cCMP	cyclic cytidine monophosphate
cDNA	complementary DNA
CFE	continuous-flow electrophoresis
cGMP	cyclic guanosine monophosphate
CHAPS	3-(3-cholamidopropyl)dimethylammonio-1-propanesulfonate
CHAPSO	3-(3-cholamidopropyl)dimethylammonio-2-hydroxy-1-propanesulfonate
CNS	central nervous system
CRF	corticotropin releasing factor
CSF-1	colony-stimulating factor-1
cUMP	cyclic uridine monophosphate
DABCO	1,4 diazabicyclo[2-2-2]octane
DDSA	dodecenyl succinic anhydride
DGDG	digalactosyldiglyceride
DHT	dihydrotestosterone
DIDS	4,4′-diisothiocyanostilbene-2,2′-disulfonic acid
DMF	dimethylformamide
DMSO	dimethylsulfoxide
DNA	deoxyribonucleic acid
DOC	deoxycholate
DPG	diphosphatidylglycerol
dsDNA	double-stranded DNA
dsRNA	double-stranded RNA
EBSS	Earle's balanced salt solution
EDTA	ethylenediaminetetraacetic acid
EGF	epidermal growth factor
EGTA	ethylene glycol-bis(β-aminoethyl ether) *N*,*N*,*N*′,*N*′-tetraacetic acid
EM	electron microscopy
ER	endoplasmic reticulum
FGF	fibroblast growth factor
FLM	frequency of labeled mitosis
FSH	follicle-stimulating hormone

GABA	γ-aminobutyric acid
G-CSF	granulocyte colony-stimulating factor
GH	growth hormone
GM-CSF	granulocyte-macrophage colony-stimulating factor
GnRH	gonadotropin releasing hormone
GRH	growth hormone releasing hormone
GRP	gastrin releasing peptide
HBSS	Hanks' balanced salt solution
hCG	human chorionic gonadotropin
Hepes	*N*-2-hydroxyethylpiperazine-*N'*-ethane-sulfonic acid
Hepps	*N*-2-hydroxyethylpiperazine-*N'*-propane-sulfonic acid
hsp-90	heat shock protein-90
5-HT	5-hydroxytryptamine
HVEM	high-voltage electron microscopy
IFN-α	interferon-α
IFN-β	interferon-β
IFN-γ	interferon-γ
Ig	immunoglobulin
IGF-I	insulin-like growth factor-I
IGF-II	insulin-like growth factor-II
IL-1	interleukin-1
IL-2	interleukin-2
IL-3	interleukin-3
IL-4	interleukin-4
IL-5	interleukin-5
IL-6	interleukin-6
IL-7	interleukin-7
INT	*p*-iodonitrotetrazolium violet
LDL	low-density lipoprotein
LH/hCG	luteinizing hormone/human chorionic gonadotrophin
LHRH	luteinizing-hormone-releasing hormone
LM	light microscopy
Lyso-PC	lysophosphatidylcholine
Lyso-PE	lysophosphatidyl-ethanolamine
mACh	muscarinic acetylcholine
M-CSF	macrophage colony-stimulating factor
Mes	2-(*N*-morpholino)ethane-sulfonic acid
MGDG	monogalactosyldiglyceride
MIS	mullerian inhibiting substance
MNA	methyl nadic anhydride
Mops	3-(*N*-morpholino)propane-sulfonic acid
NA	numerical aperture
nACh	nicotinic acetylcholine
NADH	reduced nicotinamide adenine dinucleotide
NADPH	reduced nicotinamide adenine dinucleotide phosphate
NANA	*N*-acetylneuraminic acid
NGF	nerve growth factor
NTP	nucleoside triphosphate
PBD	2-phenyl-5-(4-biphenyl)-1,3,4-oxadiazole
PBS	phosphate buffered saline
PBSA	Dulbecco's phosphate-buffered saline (solution A)
PC	phosphatidylcholine

PDGF	platelet-derived growth factor
PE	phosphatidylethanolamine
PEG	polyethylene glycol
PEI	polyethyleneimine
PG	phosphatidylglycerol
PI	phosphatidylinositol
Pipes	piperazine-*N*,*N*′-bis(2-ethane-sulfonic acid)
PMSF	phenylmethylsulfonyl fluoride
POPOP	1,4-di-(2-(5-phenyloxazolyl))-benzene
PPO	2,5-diphenyloxalone
PRL	prolactin
PS	phosphatidylserine
PVP	polyvinylpyrrolidone
RCF	relative centrifugal force
RI	refractive index
RIA	radioimmunoassay
RNA	ribonucleic acid
SHBG	sex-hormone-binding globulin
SITS	4-acetamido-4′-isothiocyanostilbene-2,2′-disulfonic acid
SM	sphinogomyelin
SRIF	somatostatin
SQDG	sulfoquinovosyldiglyceride
RIA	radioimmunoassay
ssRNA	single-stranded RNA
Taps	*N*-tris(hydroxymethyl)methyl-3-amino-propane-sulfonic acid
TAT	tyrosine aminotransferase
TEM	transmission electron miocroscopy
Tes	*N*-tris(hydroxymethyl)methyl-2-amino-ethane-sulfonic acid
TGF-α	transforming growth factor α
TGF-β	transforming growth factor β
TNF-α	cachectin
TNF-β	lymphotoxin
TRH	thyrotropin releasing hormone
V	vasopressin
VIP	vasoactive intestinal peptide
ZIO	zinc–iodine–osmium tetroxide

Amino acid	One-letter symbol	Three-letter symbol	Amino acid	One-letter symbol	Three-letter symbol
Alanine	A	Ala	Lysine	K	Lys
Arginine	R	Arg	Methionine	M	Met
Asparagine	N	Asn	Phenylalanine	F	Phe
Aspartic acid	D	Asp	Proline	P	Pro
Cysteine	C	Cys	Serine	S	Ser
Glutamine	Q	Gln	Threonine	T	Thr
Glutamic acid	E	Glu	Tryptophan	W	Trp
Histidine	H	His	Tyrosine	Y	Tyr
Isoleucine	I	Ile	Valine	V	Val
Leucine	L	Leu	Unknown or ‘other’	X	

CHAPTER 1
SOLUTIONS FOR CELL BIOLOGY

G. B. Dealtry

Solutions

Recipes for common balanced salt solutions, buffers, serum-free media and the pH and osmolarity of selected solutions are given. For more information on solutions used in tissue culture, molecular biology and plant biology refer to other books in this series.

Table 1. Balanced salt solutions

Component or substance	EBSS[b] ($g\,l^{-1}$)	PBSA[b] ($g\,l^{-1}$)	HBSS[b] ($g\,l^{-1}$)	Spinner salt solution (Eagle) ($g\,l^{-1}$)	Modified Locke (insects)[a] ($g\,l^{-1}$)
Inorganic salts					
$CaCl_2$ (anhyd.)	0.02	—	0.14	—	0.25
$CaCl_2 \cdot 2H_2O$	—	—	—	—	—
KCl	0.04	0.2	0.4	0.4	0.42
KH_2PO_4	—	0.2	0.06	—	—
$MgCl_2$ (anhyd.)[b]	—	—	—	—	—
$MgCl_2 \cdot 6H_2O$	—	—	0.1	—	—
$MgSO_4$ (anhyd.)[b]	—	—	—	—	—
$MgSO_4 \cdot 7H_2O$	0.2	—	0.1	0.2	—
NaCl	6.68	8.0	8.0	6.8	9.0
$NaHCO_3$	2.2	—	0.35	2.2	0.2
Na_2HPO_4	—	—	—	—	—
$Na_2HPO_4 \cdot 2H_2O$	—	—	—	—	—
$Na_2HPO_4 \cdot 7H_2O$	—	2.16	0.09	—	—
$NaH_2PO_4 \cdot H_2O$	0.14	—	—	1.4	—
Citric acid ($1H_2O$)	—	—	—	—	—
Na fumarate	—	—	—	—	—
Oxaloacetic acid	—	—	—	—	—
Tris(hydroxymethyl)—aminomethane	—	—	—	—	—
Other components					
D-glucose	1.0	—	1.0	1.0	2.5
Phenol red	0.01	—	0.01	0.01	—
Reference	1	1	1	1	2

Table 1. Continued

Component or substance	Holtfreter (amphibia and fish)[a] ($g\,l^{-1}$)	Frog Ringer (amphibia)[a] ($g\,l^{-1}$)	Tris–citrate buffered BSS ($g\,l^{-1}$)	Tyrode ($g\,l^{-1}$)
Inorganic salts				
$CaCl_2$ (anhyd.)	—	—	0.11	0.2
$CaCl_2 \cdot 2H_2O$	0.1	0.16	—	—
KCl	0.05	0.14	0.26	0.2
KH_2PO_4	—	—	0.04	—
$MgCl_2$ (anhyd.)[b]	—	—	—	—
$MgCl_2 \cdot 6H_2O$	—	—	—	0.1
$MgSO_4$ (anhyd.)[b]	—	—	—	—
$MgSO_4 \cdot 7H_2O$	—	0.39	2.47	—
NaCl	3.5	6.5	5.27	8.0
$NaHCO_3$	0.2	0.2	—	1.0
Na_2HPO_4	—	—	—	—
$Na_2HPO_4 \cdot 2H_2O$	—	—	0.04	—
$Na_2HPO_4 \cdot 7H_2O$	—	—	—	—
$NaH_2PO_4 \cdot H_2O$	—	—	—	0.05
Citric acid ($1H_2O$)	—	—	1.05	—
Na fumarate	—	—	0.16	—
Oxaloacetic acid	—	—	0.13	—
Tris(hydroxymethyl)-aminomethane	—	—	1.92	—
Other components				
D-glucose	—	—	1.8	1.0
Phenol red	—	—	0.02	—
Reference	2	2	2	2

NB Na fumarate and oxaloacetic acid are required only with a few cell strains at low inoculum density.

[a] Indicates defined media. These vary in complexity from the other media in that they contain a larger number of different amino acids and vitamins and are often supplemented with extra metabolites (e.g. nucleosides) and minerals. They are not used for general purposes as the other media are, but are used for many different warm-blooded animals and in tissue culture.

[b] Abbreviations: EBSS, Earle's balanced salt solution; PBSA, Dulbecco's phosphate-buffered saline (solution A); HBSS, Hanks balanced salt solution; anhyd., anhydrous.

Table 2. Amine buffers useful for biological research (3)

Chemical name	pKa at 20°C	?pKa/°C
2-(*N*-Morpholino)ethane-sulfonic acid (Mes)	6.15	−0.011
Bis(2-Hydroxyethyl)imino-tris-(hydroxymethyl)methane (Bistris)	6.5	—
N-(2-Acetamido)iminodiacetic acid (ADA)[a]	6.6	−0.011
Piperazine-*N*,*N*′-bis(2-ethane-sulfonic acid) (Pipes)	6.8	−0.0085
1,3-Bis(tris(hydroxymethyl)-methylamino)propane (Bistrispropane)	6.8(9.0)	—
N-(Acetamido)-2-aminoethane-sulfonic acid (Aces)	6.9	−0.020
3-(*N*-Morpholino)propane-sulfonic acid (Mops)	7.15	−0.013
N,*N*′-Bis(2-hydroxyethyl)-2-amino-ethane-sulfonic acid (Bes)	7.15	−0.016
N-Tris(hydroxymethyl)methyl-2-amino-ethane-sulfonic acid (Tes)	7.5	−0.020
N-2-Hydroxyethylpiperazine-*N*′-ethane-sulfonic acid (Hepes)[b]	7.55	−0.014
N-2-Hydroxyethylpiperazine-*N*′-propane-sulfonic acid (Hepps)[b]	8.1	−0.015
N-Tris(hydroxymethyl)methylglycine (Tricine)[a]	8.15	−0.021
Tris(hydroxymethyl)aminomethane (Tris)	8.3	−0.031
N,*N*-Bis(2-hydroxyethyl)glycine (Bicine)[a]	8.35	−0.018
Glycylglycine (Glycylglycine)[a]	8.4	−0.028
N-Tris(hydroxymethyl)methyl-3-amino-propane-sulfonic acid (Taps)	8.55	−0.027
1,3-Bis(tris(hydroxymethyl)-methylamino)propane (Bistrispropane)	9.0 (6.8)	—
Glycine (Glycine)[a]	9.9	—

[a] These substances may bind certain di- and polyvalent cations; therefore they may sometimes be useful for providing constant, low-level concentrations of free heavy metal ions (heavy metal buffering).
[b] These substances interfere with and preclude the Folin protein assay.

Table 3. Some buffer solutions and their pH ranges (4)

Acidic constituent	Basic constituent	pH Range
HCl	Glycine	1.0–3.7
HCl	Sodium citrate, acid	1.0–5.0
p-Toluenesulfonic acid	Sodium *p*-toluenesulfonate	1.1–3.3
Potassium bisulfosalicylate	NaOH	2.0–4.0
Phenylacetic acid; H_3BO_3; H_3PO_4	NaOH	2.0–12.0
Barbital (5,5-Diethylbarbituric acid); citric acid; H_3BO_3; KH_2PO_4	NaOH	2.0–12.0
HCl	Potassium biphthalate	2.2–4.0
Citric acid	NaOH	2.2–6.5
Citric acid	Na_2HPO_4	2.2–8.0
2-Furoic acid	Sodium furoate	2.8–4.4
Formic acid	NaOH	2.8–4.6
Succinic acid	$Na_2B_4O_7 \cdot 10H_2O$ (borax)	3.0–5.8
Phenylacetic acid	Sodium phenylacetate	3.4–5.1
Acetic acid	Sodium acetate	3.7–5.6
Potassium biphthalate	NaOH	4.0–6.2
Sodium bisuccinate	Sodium succinate	4.8–6.3
Sodium citrate, acid	NaOH	5.0–6.3
Sodium bimaleate	NaOH	5.2–6.8
KH_2PO_4	NaOH	5.8–8.0
KH_2PO_4	$Na_2B_4O_7 \cdot 10H_2O$ (borax)	5.8–9.2
NaH_2PO_4	Na_2HPO_4	5.9–8.0
HCl	Triethanolamine	6.7–8.7
HCl	Barbital sodium (sodium 5,5-diethylbarbiturate)	7.0–9.0
Barbital (5,5-diethylbarbituric acid)	Barbital sodium (sodium 5,5-diethylbarbiturate)	7.0–9.0
H_3BO_3 or HCl	$Na_2B_4O_7 \cdot 10H_2O$ (borax)	7.0–9.2
HCl	Tris(hydroxymethyl)amino-methane	7.2–9.0
H_3BO_3	NaOH	8.0–10.0
Potassium *p*-phenolsulfonate	NaOH	8.2–9.8
Glycine	NaOH	8.2–10.1
NH_4Cl	NH_4OH	8.3–9.2
Glycine; Na_2HPO_4	NaOH	8.3–11.9
HCl	Ethanolamine	8.6–10.4
$Na_2B_4O_7 \cdot 10H_2O$ (borax)	NaOH	9.2–11.0
HCl or $NaHCO_3$	Na_2CO_3	9.2–11.0
$Na_2B_4O_7 \cdot 10H_2O$ (borax)	Na_2CO_3	9.2–11.0
Na_2HPO_4	NaOH	11.0–12.0

Table 4. pH values for miscellaneous buffer solutions over a range of temperature (5–9)

Composition of the buffer solution[a]	Molarity, M[a] (mol kg^{-1})	Temperature (°C)					Ref.
		0	10	20	30	40	
Piperazine phosphate (M)	0.05	6.589	6.463	6.348	6.243	6.149	1
Hydrochloric acid (M)	0.10	7.082	6.889	6.710	6.540	6.378	2
Tris(hydroxymethyl) aminomethane (Tris) (M), Tris–HCl (M)	0.05	8.946	8.614	8.313	8.036	7.777	3
Tris (M), Tris–HCl (3M)	0.01667	8.471	8.142	7.840	7.563	7.307	4
N-Tris(hydroxy-methyl)methylglycine (Tricine) (M), Na Tricinate (M)	0.05	—	8.375	8.175	7.988	7.817	5

[a] Abbreviation: M, molarity.

Table 5. Osmolarities of some solutions and tissue culture media (10)

Media	Osmolarity (mOSm kg^{-1})
Dulbecco's phosphate-buffered saline (PBS)	294
Dulbecco's phosphate-buffered saline, without Ca^{2+} and Mg^{2+}	284
Earle's balanced salt solution (EBSS)	281
Earle's balanced salt solution, without Ca^{2+} and Mg^{2+}	287
Hanks' balanced salt solution (HBSS)	286
Hanks' balanced salt solution, without Ca^{2+} and Mg^{2+}	283
Dulbecco's modified Eagle medium	338
Eagle's basal medium, Earle's salts (EBME), with glutamine	298
Eagle's basal medium, Hanks' salts (HBME), with glutamine	290
Eagle's minimal essential medium, Earle's salts (EMEM)	297
RPMI 1640	287

REFERENCES

1. Freshney, R.I. (1983) *Culture of Animal Cells: A Manual of Basic Technique*. A. R. Liss, New York, p. 70.

2. Paul, J. (1973) *Cell and Tissue Culture (4th edn)*. Churchill Livingstone, Edinburgh.

3. Good, N. (1976) in *Handbook of Biochemistry and Molecular Biology (3rd edn), Physical and Chemical Data* (G. D. Fasman ed.). CRC Press, Ohio, Vol. I, p. 368.

4. Bates, R.G. (1972) in *Biology Data Book (2nd edn)* (P. L. Altman and D. S. Dittmer eds). FASEB Publications, Maryland, Vol. I, p. 473.

5. Hetzer, H.B., Robinson, R.A. and Bates, R.G. (1968) *Anal. Chem.*, **40**, 634.

6. Paabo, M. and Bates, R.G. (1970) *J. Phys. Chem.*, **74**, 702.

7. Bates, R.G. and Robinson, R.A. (1973) *Anal. Chem.*, **45**, 420.

8. Durst, R.A. and Staples, B.R. (1972) *Clin. Chem.*, **18**, 206.

9. Bates, R.G. Roy, R.N. and Robinson, R.A. (1973) *Anal. Chem.*, **45**, 1663.

10. Waymouth, C. (1973) in *Tissue Culture: Methods and Applications* (F. Kruse, Jr. and M. K. Patterson, Jr. eds). Academic Press, London, p. 706.

CHAPTER 2
SUBCELLULAR FRACTIONATION

J. M. Graham

Fractionation

1. INTRODUCTION

1.1. Choice of tissue or cell

For the routine production of all or selected subcellular fractions the material used most often is liver. The vast majority of fractionation methods have been worked out on the basis of this tissue. The tissue is relatively easy to work with and the ready availability of large quantities of material and the extensive literature on liver function make this the tissue of choice. Moreover, the composition and functional competence of its isolated membranes tend to be reproducible, well defined and can be related to the well-studied physiological and metabolic functions of the intact liver. Most of the methodology detailed in this chapter is based on liver. Special requirements may dictate the use of some other tissue, but compared to liver they all tend to be more difficult to homogenize and/or fractionate.

One of the most difficult membranes to obtain in a high yield and a high degree of purity is the plasma membrane. Again liver tends to be one of the most common sources, although the mammalian erythrocyte membrane offers the highest yield and purity of any preparation. The limited and specific functional capability of the erythrocyte compromises its desirability as a source material; however, because of the universality of general membrane structure, it remains a very useful membrane for structural studies.

Tissues suffer from one major drawback — the presence of a variety of cell types — which compromises the validity of observations relevant to only one of the cell types. The culture of animal cells overcomes this problem but raises others, notably the generation of large amounts of material and, except perhaps for cultured hepatocytes, a decrease in the ease and reproducibility of homogenization. For membrane fractionation, cells grown as adherent monolayers provide the best source of material, while suspension culture cells are generally more difficult to handle.

1.2. Homogenization strategy

In the case of liver and some other organized tissues such as kidney and intestinal mucosa, the homogenization procedure is relatively standard and can be applied with some degree of certainty regarding its outcome. With the exception of the erythrocyte, all other cells and tissues need to be thoroughly investigated with respect to their susceptibility to homogenization techniques and the ease of isolation of the membrane fractions so produced. The problem is particularly acute with tissue culture cells. Not only are they less susceptible to breakage by the techniques used routinely for intact tissues, their response to disruptive forces may change during continuous culture.

The plasma membrane of cultured cells poses a particular problem; unless it is stabilized, it tends to break into fragments and/or vesicles of a variety of sizes. In intact tissues, the presence of a cytoskeleton, extracellular matrix, tight junctions and/or desmosomes all lend a stability to the plasma membrane. In cultured cells, particularly in suspension culture cells, such stabilizing elements are lacking or less well developed. Whichever technique is chosen, it is essential to monitor the process and the subsequent fractionation technique to make sure that the membrane of interest and the other membrane-bound organelles are functionally and structurally intact. The aim must be to choose conditions that are the most gentle and the most reproducible.

Routinely, the homogenate is produced in 0.25 M sucrose buffered with some organic compound, and may be supplemented with ions, chelators, protease inhibitors and/or disulfide bond reducers, although hypotonic media may be used in special instances.

1.3. Fractionation strategy

In the case of the mammalian erythrocyte membrane and bacterial protoplasts, a simple centrifugation at the appropriate RCF will separate the membrane from the only contaminant — soluble proteins. Further purification involves merely washing to release any adsorbed protein. The extent of the washing is the only problem: this arises from the question as to what fraction of the protein removed by washing is truly a contaminant and what is an extrinsically bound protein (see Chapter 6).

The standard procedure for the majority of material produced from eucaryotic cells is to produce a series of pellets by differential centrifugation and then to use gradients (often of sucrose, but increasingly of the newer, low osmolarity media, such as metrizamide, Nycodenz™, or Percoll™) to separate the particles on the basis of their buoyant (isopycnic) density, but sometimes on the basis of their sedimentation rate.

Alternatively, the membranes may be separated on the basis of their surface charge density by continuous-flow electrophoresis or by aqueous polymer partitioning (the hydrophobicity of the membrane may play some part in partitioning). Specific membranes can be isolated by using immunoadsorption to a particle which is covalently bound to an antibody raised to a membrane antigen or to a lectin.

It is necessary to monitor the efficacy of the fractionation process by assaying fractions for membrane markers (normally enzymes), so as to assess the purity of the recovered product, its recovery and the degree of contamination by other membranes.

2. HOMOGENIZATION

2.1. Types of homogenizer

Continuous and instantaneous

In the continuous variety, the material suspension is subject to the disruptive force for a period which is defined by the operator: some cells will be disrupted during the first few seconds of the process, the rest at later times. Consequently, the released organelles will also be exposed to these forces for a variable time. Most mechanical and liquid shear devices are of this type.

In the instantaneous variety, the material is subjected to the disruptive force once only. Gaseous shear (nitrogen cavitation) and pressure extrusion devices are of this type.

Commercially available homogenizers

Table 1 and *Table 2* list the different types of homogenizer that are currently available and their routine uses.

Rotating blade homogenizers. These can deal with a large range of sample types and sizes (see *Tables 1* and *2*). Their main use is for plant tissues, animal tissues that tend to be fibrous, or

► p. 10

Table 1. Types and volumes of homogenizer

Type	Varieties	Volumes
Rotating blades, motor beneath work head	Domestic liquidizer Waring blender	> 200 ml > 200 ml
Rotating blades, motor above work head	Domestic hand-held liquidizer Ultra-Turrax Polytron	100 ml–1 liter 20 ml–1 liter 2–100 ml
Hand-operated liquid shear	Dounce	1–50 ml
Mechanical liquid shear	Potter–Elvehjem	1–50 ml
Gaseous shear (nitrogen cavitation)		10–500 ml
Pressure extrusion	French press	5–50 ml
Sonication		5–200 ml

Table 2. Homogenizers — uses and variable parameters

Type	Tissue	Variables
Rotating blades	Plant tissues Hard animal tissue (muscle) Large amount of soft animal tissue (e.g. beef heart)	Speed of rotation Presence of abrasive particles (e.g. glass beads)
Dounce	Soft animal tissue Tissue culture cells	Pestle clearance Thrust of pestle Number of passes
Potter–Elvehjem	Soft animal tissue	Pestle clearance Speed of rotation Number of passes
Nitrogen cavitation	Tissue culture cells	Nitrogen pressure Equilibration time
Pressure extrusion	Bacteria Tissue culture cells	Pressure Orifice diameter
Sonication	Bacteria	Wattage Sonication time

for large amounts of any tissue. The severity of the disruption is determined by the time, speed of rotation of the blades and, to some extent, by the configuration of the blades themselves. They are not normally used for the production of organelles from tissues such as liver, but may be used in special cases, for example the isolation of mitochondria from beef heart.

Liquid shear homogenizer (hand held). The Dounce homogenizer is used routinely for this task; its pestle is a glass ball and the clearance between the ball and the wall of the containing vessel is often imprecisely referred to as loose-fitting or tight-fitting (see *Tables 1* and *2*). Tight-fitting models are used for tissue culture cells, and loose-fitting ones for gentle homogenization of soft tissues such as liver. Normally, between 15 and 20 strokes of the pestle are required in both cases for complete homogenization. Potter–Elvehjem homogenizers can be used in a hand-operated mode but the results tend to be difficult to reproduce.

Liquid-shear (motor driven). The Potter-Elvehjem homogenizer (Teflon pestle) attached to a thyristor-controlled electric motor is normally used. The clearances routinely vary from 0.05 to 0.5 mm; speeds of rotation are commonly about 500 r.p.m. and between 4 and 20 passes of the pestle are required. Fibrous tissue should not be used with such an homogenizer and, generally, tissue culture cells are not effectively ruptured.

French pressure cell. Pressures of 100–150 MPa are commonly employed, sometimes the process is repeated.

Nitrogen cavitation. Pressures of about 5 MPa (50 bar) for an equilibration time of 15 min are used. Unlike the French pressure cell, the high pressure is used to 'dissolve' the nitrogen in the cytosol and rupture occurs when the pressure is suddenly released. In the French pressure cell, the pressure is used to force the material through a small orifice.

Sonication. Commonly used conditions are a power of 150 W, with a probe of 0.5–1.0 cm diameter, using 15 sec bursts with 15 sec cooling intervals in between. It is probably the most disruptive of techniques and is rarely used for membrane fractionation from eucaryotes because all organelles tend to form vesicles and to release their contents, at least in part —this is a particular problem with nuclei since the released DNA causes extensive aggregation.

2.2. Tissues

Liver is the most widely used tissue for membrane preparation, and the type of homogenizer and the medium used are dictated largely by the organelle which is to be purified. *Table 3* lists some of the more commonly isolated fractions and the homogenization conditions. Generally, the clearance of the Potter–Elvehjem homogenizer should be about 0.09 mm and complete disruption (greater than 95%) should be achieved after 5–8 up-and-down strokes of the pestle. Disruption of liver by Dounce homogenization normally requires 15–20 strokes with a loose-fitting pestle. See *Table 3* for references.

Table 4 gives the homogenization conditions used for some of the other tissues that may be encountered.

2.3. Tissue culture cells

General approach and problems

Depending on the cell type, it may be necessary to stress the cells osmotically before they can be disrupted reproducibly by Dounce homogenization. Generally speaking, Potter–Elvehjem homogenization is not very effective. Osmotic stress can be accomplished by suspending the cells in a grossly hypo-osmotic medium. The osmotic gradient across the plasma membrane can be enhanced by first increasing the osmolarity of the cytosol by allowing the cells to take up glycerol.

Table 3. Homogenizer and medium for disruption of cells and tissues and isolation of organelles from soft animal tissue (e.g. liver)

Organelle	Homogenization medium	Homogenizer
Nuclei (1, 2)[a]	0.25 M sucrose, 25 mM KCl, 5 mM $MgCl_2$, 50 mM Tris–HCl, pH 7.6	Potter–Elvehjem
Mitochondria (3)	0.2 M mannitol, 0.05 M sucrose, 1 mM EDTA,[b] 10 mM KCl, 10 mM Hepes–NaOH, pH 7.5	Potter–Elvehjem
Plasma membrane sheets (4, 5)	1 mM $NaHCO_3$	Dounce
Lysosomes, Golgi, microsomes, etc. (3)	0.25 M sucrose, 10 mM Hepes–NaOH, pH 7.6	Potter–Elvehjem

[a]If the nuclei are going to be used for a preparation of nuclear membrane by a high ionic strength method, include 1 mM EGTA (ethylene glycol-bis (β-aminoethyl ether)N,N,N',N'-tetraacetic acid) and 1 mM PMSF (phenylmethylsulfonyl fluoride) in the medium.
[b]EDTA = ethylenediaminetetraacetic acid.

Table 4. Homogenization of other material

Material	Homogenizer	Homogenization medium	Comments
Muscle (6, 7)	Waring blender or similar device	0.12 M NaCl, 5 mM Tris–HCl, pH 7.6	For sarcoplasmic reticulum
Plants (8, 9)	Waring blender or similar device	330 mM sorbitol, 10 mM pyrophosphate, 5 mM $MgCl_2$, 2 mM Na-iso-ascorbate, pH 6.5	For chloroplasts
		0.3 M mannitol, 2 mM EDTA, 0.1% BSA,[b] 40 mM Mops,[c] pH 7.5, 0.4–0.6% polyvinyl-pyrrolidone	For mitchondria
Bacteria[a] (10, 11)	Sonicator or French pressure cell	2 mM $MgCl_2$, 1 mM EGTA, 50 mM Tris–HCl, pH 7.4	Produces membrane fragments and vesicles

[a]Spheroplasts can be produced from bacteria by enzymic lysis using lysozyme and a cocktail of enzymes for digesting nucleic acids (12).
[b]BSA = bovine serum albumin.
[c]Mops = 3-(N-morpholino)propane-sulfonic acid.

Fractionation

Iso-osmotic sucrose (buffered with triethanolamine/acetic acid) can be used for adherent monolayer cells: alternatively, nitrogen cavitation (or possibly pressure extrusion) can be effective in an iso-osmotic medium.

The major problem with using osmotic stress is that tissue culture cell nuclei tend to be very fragile, and when released in the absence of an osmotic balancer and cations they swell and release their DNA. On the other hand, the presence of these solutes makes the plasma membrane more resilient to rupture, in which case sucrose, $MgCl_2$ and/or $CaCl_2$ and/or KCl can be added immediately after homogenization.

In all instances, use a tight-fitting Dounce homogenizer, except where indicated. Where a hypo-osmotic medium is used, make the cell number to homogenization medium volume as high as possible so that the released cytoplasmic proteins can exert a maximum protective effect on the organelles. If the method requires more than 25 passes of the pestle, use an alternative technique or modify the composition of the homogenization medium.

Some of the commonly used media are given in *Table 5*.

Glycerol loading

The method was developed primarily for blood platelets (18) but it can be adapted to a number of cell types (19, 20). The incubation is performed at room temperature in buffered glycerol (1.2–4.0 M) for 10 min. Then the cells are harvested and resuspended in buffered 0.1–0.25 M sucrose at 0°C for 5 min.

If the cells swell but do not rupture, an additional shearing force (vortex mixing, pipetting, or Dounce homogenizer) can be applied.

Table 5. Homogenization media for tissue culture cells

Cell suspension medium	Permitted supplements	Comments
0.25 M sucrose, 10 mM triethanolamine, 10 mM acetic acid, pH 7.6 (13)	1 mM EDTA	Best for monolayer cells
1 mM $NaHCO_3$, pH 8.0 (14)	<2 mM divalent cations	Effective for suspension culture cells, mainly for plasma membrane preparation
10 mM imidazole–HCl 10 mM KCl, pH 7.6 (15)	<2 mM divalent cations	
0.25 M sucrose, 10 mM Hepes–NaOH, pH 7.6 (16, 17)	Any	Use nitrogen cavitation at 5 MPa for 15 min or pressure extrusion

3. ISOLATION OF SUBCELLULAR FRACTIONS

3.1. Centrifugation

Sedimentation properties of particles

The sedimentation properties of a subcellular particle in a low-density medium such as 0.25 M sucrose depend largely on the size of the particle; only in the case of the nucleus does its density also play a significant part.

Table 6 gives the range of sizes that can be expected for most biological particles and the centrifugal force (g_{av}) needed to pellet the particle from a 50 ml suspension in 0.25 M sucrose. For larger volumes, which occupy centrifugation bottles with a longer sedimentation

Table 6. Size and sedimentation properties of subcellular particles

Subcellular particle	Size (μm)	RCF (g_{av})[a]	Time (min)
Nucleus	4–12	500–1000	5–10
Nuclear membrane[b]		2000 (30 000)	30 (5)
Mitochondria	0.4–2.5	1000–10 000	10–15
Lysosomes	0.4–0.8	6000–15 000	10–20
Peroxisomes	0.4–0.8	6000–15 000	10–20
Rough ER vesicle[c]	0.05–0.35	30 000–100 000	30–60
Smooth ER vesicle[c]	0.05–0.3	50 000–100 000	30–60
Plasma membrane			
sheet	3–20	1000–3000	10–15
vesicles	0.05–2.0	50 000–100 000	30–60
Endosome	0.05–0.4	50 000–100 000	30–60
Golgi (intact)	1.0–2.0	10 000–20 000	20–30
Golgi (vesicle)	0.05–0.5	50 000–100 000	20–40
Sarcoplasmic reticulum	0.1–1.0	10 000–35 000	20
Chloroplasts	2–5	1000–2000	10
Plant mitochondria	1–3	5000–20 000	15

[a] RCF = relative centrifugal force.
[b] Nuclear membrane preparations are intact envelopes (ghosts), whose sedimentation rate depends on the mode of preparation.
[c] ER = endoplasmic reticulum.

pathlength, it should be borne in mind that those particles nearer the axis of rotation may not experience a high enough centrifugal force to sediment satisfactorily in the given time.

Purification of subcellular particles in gradients

Although sucrose gradients are commonly used for the purification of subcellular particles, they do suffer from a number of disadvantages. Since the concentrations used are frequently in the range 20–50% (w/w), these gradients are hyperosmotic throughout; consequently, osmotically active organelles lose water from their internal compartments and increase in density as they travel down through the gradient. If they are susequently suspended in an iso-osmotic medium, and if they are then subjected to further gradient centrifugation, as is often the case, such organelles will undergo repeated shrinking and swelling.

Table 7. Density of subcellular particles in sucrose and low osmolarity media[a]

Subcellular particle	**Density ($g\ cm^{-3}$)**	
	Sucrose	**Low osmolarity medium**[a]
Nucleus	> 1.30	1.21–1.24
Nuclear membrane	1.18–1.22	na
Mitochondria	1.17–1.21	1.15–1.20
Mitoplasts	1.14	na
Lysosomes	1.19–1.21	1.10–1.15
Peroxisomes	1.18–1.23	1.19–1.22
Rough ER	1.18–1.26	na
Smooth ER	1.06–1.15	1.03–1.07
Plasma membrane		
sheets	1.14–1.19	1.12–1.15
vesicles	1.07–1.16	1.05–1.12
Endosomes	1.06–1.16	1.05–1.10
Golgi (intact)	1.05–1.12	1.03–1.08
Golgi (vesicle)	1.05–1.12	1.03–1.08
Sarcoplasmic reticulum	1.04–1.08	na
Chloroplasts	1.18–1.20	1.10–1.13
Plant mitochondria	1.16–1.19	na

[a] These figures are densities in Nycodenz™, metrizamide, or Percoll™.
na = not available.

The change in density undergone by particles in sucrose gradients can compromise resolution. In such gradients particles such as mitochondria, lysosomes and peroxisomes, for example, exhibit densities that overlap significantly. Separation of components on the basis of size can also be less than ideal if shrinkage occurs during centrifugation. Loss of water from internal compartments may also be deleterious as regards to the function of the organelles.

The use of Nycodenz™ (or metrizamide) offers much lower osmolarities than sucrose for gradients in the same density range (21, 22). By judicious selection of co-solutes Nycodenz™ gradients can be made virtually iso-osmotic throughout (23). Percoll™, being a colloidal solution, has essentially no osmotic effect, so gradients of this compound are made iso-osmotic even more easily, by addition of an appropriate solute (24). On the other hand, since Percoll™ particles themselves sediment fairly rapidly, the use of Percoll™ gradients at centrifugal forces above 20 000 g_{av} is very difficult.

Almost without exception, organelles have lower densities in these low osmolarity media than they do in sucrose (see *Table 7*), but they will also be larger since they do not lose water.

The other significant difference between sucrose and Nycodenz™ (or Percoll™) is the much lower viscosity of the low osmolarity media. The effect of this is to reduce the times required for particles to sediment (or float up) through Nycodenz™ or Percoll™ gradients. The use of a low-viscosity medium also means that the harvesting of gradient fractions by centrifugation only requires the medium to be diluted by 50% with buffer, rather than by the 2–3 volumes that are normally used for sucrose gradients.

Table 8. Definition of differential centrifugation pellets (liver)

Pellet	**RCF × time**	**Content**[a]
P1	1000 g_{av} for 10 min	Nuclei, heavy mitochondria, plasma membrane sheets
P2	3000 g_{av} for 10 min	Heavy mitochondria, plasma membrane fragments
P3	6000 g_{av} for 10 min	Mitochondria, lysosomes, peroxisomes, intact Golgi
P4	10 000 g_{av} for 10 min	Mitochondria, lysosomes, peroxisomes, Golgi membranes
P5	20 000 g_{av} for 10 min	Lysosomes, peroxisomes, Golgi membranes, large and dense vesicles (e.g. rough ER)
P6	100 000 g_{av} for 50 min	All vesicles from ER, plasma membrane, Golgi, endosomes, etc.

[a] In practice the pellets will show more cross-contamination of particles than suggested. Slowly sedimenting material may co-sediment with, by entrapment in, large numbers of more rapidly sedimenting particles; a problem which is overcome to some extent by washing the pellets. Plasma membrane fragments may be of a variety of sizes which sediment at all speeds. The majority of mitochondria will be in P2 and P3, the majority of lysosomes and peroxisomes in P4. Golgi membranes will only occur in P3 if they are intact.

Table 9. Recommended gradient and centrifugation conditions for purification of subcellular particles

Particle[a]	Pellet[a]	RCF (g_{av}) × time	Gradient[a]	Contaminants[a]	Ref.
NUC	P1	100 000 g × 60 min	60% sucrose barrier	none	1, 25
	P1/HOM	15 000 g × 60 min	20–50% dc Nycodenz™	none	26
PMS	P1	160 000 g × 3 h	37%/60% sucrose	MIT	3
MIT	P2	3000 g × 10 min	none	NUC	
	P2 + P3 + P4	50 000 g × 2 h	20–40% dc Nycodenz™	PER	26
	HOM	37 000 g × 30 sec	20–52% dc Percoll™	PER	3
LYS	P4 + P5	50 000 g × 2 h	10–50% dc Nycodenz™	PM	26
	P4 + P5	37 000 g × 30 min	15–85% c Percoll™	MIT	3
PER	P4 + P5	95 000 g × 2 h	34–47% dc Nycodenz™	MIT	27
GOLGI	P4 + P5	160 000 g × 1 h	10–44% dc sucrose	PMV/SER	3
	P4 + P5	50 000 g × 2 h	10–50% dc Nycodenz™	LYS	26
SER/RER	P6	150 000 g × 1 h	20%/45% sucrose with 15 mM CsCl	PMV/END	28
PMV[b]	P4 + P5/P6	100 000 g × 4 h	10–30% c metrizamide	SER/END	29
END[c]	HOM/P6	85 000 g × 45 min	5–35% c Nycodenz™	SER/PMV	30

Table 9. Continued

Particle[a]	Pellet	RCF (g_{av}) × time	Gradient[a]	Contaminants[a]	Ref.
SARCRT	P2	120 000 g × 4 h	20–38% dc sucrose	MIT	7
NUCMB[d]	P1	100 000 g × 1 h	10–55% dc sucrose	none	26, 31
OUTER MITMB[d]	P2	115 000 g × 1 h	23–43% dc sucrose	INNER MITMB	32

[a]Abbreviations: NUC, nuclei; PMS, plasma membrane sheets; MIT, mitochondria; LYS, lysosomes; PER, peroxisomes; PMV, plasma membrane vesicles; SER, smooth endoplasmic reticulum; RER, rough endoplasmic reticulum; END, endosomes; SARCRT, sarcoplasmic reticulum; NUCMB, nuclear membrane; OUTER MITMB, outer mitochondrial membrane; INNER MITMB, inner mitochondrial membrane; HOM, homogenate; dc, discontinuous; c, continuous.
[b]Method can be used for the fractionation of secretory vesicles.
[c]Method can be used for the fractionation of any vesicle on the basis of size.
[d]Pellets require pretreatment before application to the gradient (see refs 26, 31, 32).

Conditions for the isolation of organelles

After disruption of the cellular material, the homogenate should be centrifuged differentially using a series of spins of increasing centrifugal forces; the exact conditions are determined by the operator as the optimum for producing a fraction of the membrane of interest in the highest yield can differ. For convenience, a standard differential centrifugation scheme is given in *Table 8*, which defines the series of pellets. A pellet definition of P4 will assume that the more rapidly sedimenting particles in P1, P2 and P3 have been discarded. A pellet definition of P2 + P3 means that P1 has been discarded and the material to be used for gradient purification is the P1 supernatant centrifuged at $6000g_{av}$ for 10 min (i.e. omitting the P2 step).

Table 9 gives the centrifugation conditions and recommended gradients for the purification of a number of subcellular particles. The starting material is usually a pellet produced by differential centrifugation of an homogenate. Except in the case of the sarcoplasmic reticulum, the material is derived from a liver homogenate produced in the manner described in *Table 4*.

3.2. Electrophoresis

Principles

The separation of particles on the basis of their surface charge is most commonly achieved by continuous-flow electrophoresis (CFE), in which the sample is injected into a vertically moving curtain of buffer across which an electric field is imposed. Biological particles are deflected towards the anode to an extent that is proportional to their surface charge density. The size and density of the particles are not important.

CFE is used as an alternative to centrifugation only in the final purification of a subcellular fraction, not in the early stages of purification of crude material. The media are normally iso-osmotic and buffered with triethanolamine/acetic acid. Surface charges on membranes can be manipulated by modulation of pH, and supplementing the medium with cations or chelating agents. The imposed electric field depends on the medium used and the specifications of the apparatus, but under normal conditions the voltage is about 1000 V (120–145 V cm^{-1}) and buffer flow rates can vary from 50 to 750 ml h^{-1}. As the eluate from the chamber is divided into some 90 fractions, this flow rate is often expressed as ml h^{-1} $fraction^{-1}$.

Separation conditions

Table 10 lists some of the membrane fractions that have been separated by CFE. In practice, it seems that to achieve useful separations, except in the case of the separation of inside-out and right-side out membrane vesicles from the human erythrocyte membrane (33), the surface charge on subcellular organelles needs to be modulated selectively by the activity of certain hydrolytic enzymes. The enzyme treatment is used to modify the surface of specific membranes and thus shift them towards either the anode or cathode, away from the bulk of the membrane material.

The separation buffers for all the systems given in *Table 10* contain 10 mM triethanolamine/10 mM acetic acid at pHs between 6.5 and 7.4; the osmotic balancer is normally 0.25 M sucrose, except in system 2 which uses sorbitol. The N-tosyl-L-phenylalanine chloromethylketone-trypsin can be used differentially: at 0.2% the lysosomes are shifted towards the anode, while at concentrations greater than 1% the endosomes are shifted. In all cases the applied voltage is of the order of 120–140 V cm^{-1} and the buffer flow rate is 2–3 ml h^{-1} $fraction^{-1}$.

Table 10. Separation of particles by continuous-flow electrophoresis

Source material	Separates	Ref.
1. Rat-liver Golgi fraction (α-amylase treated)	Mid-region cisternae	34
2. Human platelets (neuraminidase treated)	Surface and intracellular membrane domains	35
3. Fibroblast mitochondrial fraction (washed)	Lysosomes	36
4. Fibroblast post-nuclear fraction (treated with *N*-tosyl-L-phenylalanine chloromethylketone-trypsin)	Lysosomes and endosomes	37

Fractionation

3.3. Aqueous two-phase partitioning

Principles

The standard two-polymer system comprises polyethylene glycol (PEG) and dextran: when these two polymers are mixed at certain minimum concentrations they form two phases. Commonly, the concentrations used are 5% (w/w) dextran and 3.5% (w/w) polyethylene glycol but, depending on the particular separation, may be as high as 6.8% of each (37).

Partitioning of membrane organelles depends on their surface charge and their hydrophobicity, and is consequently modulated by polymer concentration and salt concentration (37). Both phases can be made iso-osmotic by the addition of an inert solute (sucrose, mannitol, etc.).

If an ion of an added salt partitions differentially between the two phases (i.e. an interfacial potential exists), this can influence the distribution of organelles. Chloride partitions equally between dextran and polyethylene glycol, while phosphate predominates in the dextran.

In affinity partitioning, a suitable ligand is covalently bound to one of the polymers: this will then influence the distribution of an organelle bearing an appropriate receptor.

Normally, multiple partitioning is required to effect an adequate resolution of the particles of interest. For a small number of transfers this can be achieved manually by mixing the sample with the polymers in an appropriate buffer, centrifuging at low speed to allow the two phases to separate, removing the upper phase to another tube and mixing it and the remaining lower phase with new lower and upper phase, respectively. Automatically this can be achieved with a counter-current distribution apparatus, which requires a single centrifugation (42).

Separations

The technique has not achieved widespread popularity, but *Table 11* describes some of the experimental conditions that have been used, and the organelles that have been fractionated. Separation 1 can resolve right-side-out from inside-out vesicles, and in separations 2 and 3

Table 11. Phase partition separations

Membrane	Dextran/PEG concentration[a]	Additions	Ref.
1. Erythrocyte membrane vesicles	5%/4%	0.09 M phosphate, pH 6.8 0.02 M NaCl	38
2. Rat-liver plasma membrane	5%/3.8%	Phosphate buffer, pH 7.0	39
3. Crayfish neural plasma membrane	5%/4%	Phosphate buffer, pH 5.5 NaCl	40
4. *Acholeoplasma* surface membrane	5%/4%	Phosphate buffer, pH 7.4 NaCl	41
5. Rat-liver organelles	5%/4.8%	Tris–maleate, pH 7.4 0–48 mM NaCl	42
6. Rat-liver organelles	5%/3.8%	10 mM phosphate, pH 7.4 0.25 M sucrose	43
7. Rat-liver microsomes	5%/3.5%	50 mM phosphate, pH 6.8 3 mM $MgCl_2$	44
8. Thylakoid membranes	5.7%/5.7%	10 mM phosphate, pH 7.4 5 mM NaCl/20 mM sucrose	45

[a] Polymer concentrations given as % w/w.

the plasma membrane is located at the interface. Separation 2 uses a low-speed nuclear pellet as the source material, and separation 3 a mitochondrial + microsomal fraction.

In separation 4, the system was able to resolve a number of membrane subfractions, and the distribution of these subfractions between the top phase, interphase and lower phase depended on the lipid type.

In separation 5 the efficiency of separation of Golgi, smooth and rough microsomes, plasma membrane and lysosomes depended on the polymer concentration (6–6.8%) and the NaCl concentration. Subfractionation of some vesicle populations was obtained in separation 7, and separation 8 achieved separation of inside-out and right-side-out vesicles, as well as separation of PSI and PSII photosystems.

4. IDENTIFICATION OF MEMBRANES

The identification of membranes and measurements of their specific activity, recovery and contamination rely largely on their enzyme profiles. Most of these profiles have come from work on liver. Their extrapolation to other soft animal tissues is more-or-less valid, but to other tissues and to tissue culture cells is less certain.

The plasma membrane from organized tissues (e.g. liver, kidney and intestinal mucosa) is organized into functionally specific domains, which are characterized by different sets of enzymes (*Table 12*). In most such cells, just two domains are recognized, an apical one which possesses microvilli, and a basolateral domain which does not. In intestinal mucosa cells (enterocytes) the apical membrane is the luminal brush border and the basolateral the remainder. The liver plasma membrane is unusually distinguished into three domains: the apical is equivalent to the bile canalicular membrane, while the basolateral domain is divided into the blood sinusoidal membrane, and the remainder is the contiguous membrane. Such domains may exist in other cell types which do not have a clear functional polarity, but their demonstration is much less clear. For a general discussion of the use of enzymes as markers see ref. 64.

Table 12. Major enzyme characteristics of mammalian cell membranes

Membrane	Enzyme	Comment
Plasma membrane		
Apical	5′-nucleotidase (46, 47) Leucine aminopeptidase (48) Alkaline phosphatase (49)	In some cells this membrane may be associated with a tissue-specific function, e.g. sugar transport in enterocytes
Basolateral	Na^+/K^+-ATPase (47, 50) Hormone-stimulated adenylate cyclase (51, 52)	In liver the Na^+/K^+-ATPase is located primarily in the contiguous membrane and the cyclase in the blood sinusoidal membrane
General	5′-nucleotidase Na^+/K^+-ATPase	These most widely used general markers are not always easy to analyze in culture cell membranes
Mitochondria	Succinate dehydrogenase (53, 54)	Can be assayed with cytochrome *c* or INT[a] as electron acceptor; levels in cultured cells tend to be low
Lysosomes	β-Galactosidase (55) Acid phosphatase (49) Aryl sulfatase (56)	
Peroxisomes	Catalase (57)	
Golgi	Galactosyl transferase (58, 59)	Can be measured using *N*-acetyl-glucosamine or ovalbumin acceptor

Table 12. Continued

Membrane	Enzyme	Comment
Endoplasmic reticulum	NADPH (or NADH)-cytochrome *c* reductase (60)[b] Glucose-6-phosphatase (61)	Only liver and kidney have significant levels of the glucose-6-phosphatase
Nuclear membrane	Nucleoside triphosphatase (62)	Contamination by ER is difficult to assess because there is some continuity between the membranes
Outer mitochondrial membrane	Monoamine oxidase (63)	
Inner mitochondrial membrane	see Mitochondria	

[a]INT, *p*-iodonitrotetrazolium violet.
[b]NADPH, reduced β-nicotinamide adenine dinucleotide phosphate; NADH, reduced β-nicotinamide adenine dinucleotide.

5. REFERENCES

1. Blobel, G. and Potter, V.R. (1966) *Science*, **154**, 1662.

2. Harris, J.R. and Agutter, P.S. (1970) *J. Ultrastr. Res.*, **33**, 219.

3. Graham J.M. (1984) in *Centrifugation — A Practical Approach (2nd edn)* (D. Rickwood, ed.) IRL Press at Oxford University Press, Oxford, p. 161.

4. Neville, D.M. (1968) *Biochim. Biophys. Acta*, **154**, 540.

5. Wisher, M.H. and Evans, W.H. (1975) *Biochem. J.*, **146**, 375.

6. MacLennan, D.H. (1970) *J. Biol. Chem.*, **245**, 4508.

7. Meissner, G. (1984) *J. Biol. Chem.*, **259**, 2365.

8. Leegood, R.C. and Walker, D.A. (1983) in *Isolation of Membranes and Organelles from Plant Cells* (J.L. Hall and A.L. Moore, eds). Academic Press, New York, p. 183.

9. Walker, D.A. (1980) *Methods Enzymol.*, **69**, 94.

10. Poole, R.K. and Haddock, B.A. (1974) *Biochem. J.*, **144**, 77.

11. Coakley, W.T., Bater, A.J. and Lloyd, D. (1977) *Adv. Microbiol. Physiol.*, **16**, 279.

12. Kaback, H.R. (1971) *Methods Enzymol.*, **22**, 99.

13. Marsh, M., Schmid, S., Kern, H., Harms, E., Male, P., Mellman, I. and Helenius, A. (1987) *J. Cell Biol.*, **104**, 875.

14. Graham, J.M. and Coffey, K.H.M. (1979) *Biochem. J.*, **182**, 173.

15. Graham, J.M. (1983) in *Iodinated Density Gradient Media — A Practical Approach* (D. Rickwood, ed.). IRL Press at Oxford University Press, Oxford, p. 91.

16. Graham, J.M. (1972) *Biochem. J.*, **130**, 1113.

17. Graham J.M. and Hynes, R.O. (1975) *Biochem. Soc. Trans.*, **3**, 761.

18. Barber, A.J. and Jamieson, G.A. (1970) *J. Biol. Chem.*, **245**, 6357.

19. Graham, J.M. and Sandall, J.K. (1979) *Biochem. J.*, **182**, 157.

20. Graham, J.M. and Coffey, K.H.M. (1979) *Biochem. J.*, **182**, 165.

21. Rickwood, D., Ford, T.C. and Graham, J.M. (1982) *Anal. Biochem.*, **123**, 23.

22. Rickwood, D. (1983) in *Iodinated Density Gradient Media — A Practical Approach* (D.

Rickwood, ed.). IRL Press at Oxford University Press, Oxford, p. 1.

23. Boyum, A., Berg, T. and Blomhoff, R. (1983) in *Iodinated Density Gradient Media — A Practical Approach* (D. Rickwood, ed.). IRL Press at Oxford University Press, Oxford, p. 147.

24. Rickwood, D. (1984) *Centrifugation — A Practical Approach (2nd edn)*. IRL Press at Oxford University Press, Oxford, p. 1.

25. Comerford, S.A., McLuckie, I.F., Gorman, M., Scott, K.A. and Agutter, P.S. (1985) *Biochem. J.*, **226**, 95.

26. Graham, J.M., Ford, T.C. and Rickwood, D. (1990) *Anal. Biochem.*, **187**, 318.

27. Wattiaux, R. and Wattiaux-De Coninck, S. (1983) in *Iodinated Density Gradient Media — A Practical Approach* (D. Rickwood, ed.). IRL Press at Oxford University Press, Oxford, p. 119.

28. Bergstrand, A. and Dallner, G. (1969) *Anal. Biochem.*, **29**, 351.

29. Graham, J.M. and Winterbourne, D.J. (1988) *Biochem. J.*, **252**, 437.

30. Kindberg, G.M., Ford, T.C., Rickwood, D. and Berg, T. (1984) *Anal. Biochem.*, **142**, 455.

31. Kay, R.R., Fraser, D. and Johnston, I.R. (1972) *Eur. J. Biochem.*, **30**, 145.

32. Coty, W.A., Wehrle, J.P. and Pedersen, P.L. (1979) *Methods Enzymol.*, **56**, 353.

33. Heidrich, H.G and Leutner, G. (1974) *Eur. J. Biochem.*, **41**, 37.

34. Navas, P., Minnifield, N., Sun, I. and Morré, D.J. (1986) *Biochim. Biophys. Acta*, **881**, 1.

35. Menashi, S., Weintraub, H. and Crawford, N. (1981) *J. Biol. Chem.*, **256**, 4095.

36. Harms, E., Kern, H. and Schneider, J.A. (1980) *Proc. Natl. Acad. Sci. USA*, **77**, 6139.

37. Fisher, D. (1981) *Biochem. J.*, **196**, 1.

38. Walter, H. and Krob, E.J. (1976) *Biochim. Biophys. Acta*, **455**, 8.

39. Lesko, L., Donlon, M., Marinetti, G.V. and Hare, J.D. (1973) *Biochim. Biophys. Acta*, **311**, 173.

40. Uehara, S. and Uyemura, K. (1979) *Biochim. Biophys. Acta*, **556**, 96.

41. Weislander, A.K.E., Christiansson, A., Walter, H. and Weibull, C. (1979) *Biochim. Biophys. Acta*, **550**, 1.

42. Hino, Y., Asano, A. and Sato, R. (1978) *J. Biochem.*, **83**, 925.

43. Morris, W.B. and Peters, T.J. (1980) *Biochem. Soc. Trans.*, **8**, 76.

44. Ohlson, R., Jergil, B. and Walter, H. (1978) *Biochem. J.*, **172**, 189.

45. Akerlund, H.-E. and Andersson, B. (1983) *Biochim. Biophys. Acta*, **725**, 34.

46. Widnell, C.C. (1974) *Methods Enzymol.*, **32**, 368.

47. Avruch, J. and Wallach, D.F.H. (1972) *Biochim. Biophys. Acta*, **233**, 334.

48. Wachsmuth, E.D., Fritz, E. and Pfleiderer, G. (1966) *Biochemistry*, **5**, 169.

49. Engstrom, L. (1961) *Biochim. Biophys. Acta*, **52**, 36.

50. Esmann, M. (1988) *Methods Enzymol.*, **156**, 103.

51. Shima, S., Komoriyama, K., Hirai, M. and Kouyama, H. (1983) *J. Biol. Chem.*, **258**, 2083.

52. Gilman, A.G. (1970) *Proc. Natl. Acad. Sci. USA*, **67**, 305.

53. Mackler, B., Collip, P.J., Duncan, H.M., Rao, N.A. and Heunnekens, F.M. (1962) *J. Biol. Chem.*, **237**, 2968.

54. Butcher, R.G. (1970) *Exp. Cell Res.*, **60**, 54.

55. Vaes, G. (1966) *Methods Enzymol.*, **8**, 509.

56. Chang, P.L., Rosa, N.E. and Davidson, E.G. (1981) *Anal. Biochem.*, **117**, 382.

57. Cohen, C., Dembiec, D. and Marcus, J. (1970) *Anal. Biochem.*, **34**, 30.

58. Fleischer, B., Fleischer, S. and Ozawa, H. (1969) *J. Cell Biol.*, **43**, 59.

59. Beaufay, H., Amar-Costesec, A., Feytmans, E., Thines-Sempoux, D., Wibo, M., Robbi, M. and Berthet, J. (1974) *J. Cell Biol.*, **61**, 188.

60. Williams, C.H. and Kamin, H. (1962) *J. Biol. Chem.*, **237**, 587.

61. Aronson, N.N. and Touster, O. (1974) *Methods Enzymol.*, **31**, 90.

62. Agutter, P.S., Cockrill, J.B., Lavine, J.E., McCaldin, B. and Sims, R.B. (1979) *Biochem. J.*, **181**, 647.

63. Williams, C.H., Lawson, J. and Backwell, F.R.C. (1988) *Biochem. J.*, **256**, 911.

64. Graham J.M. (1975) in *New Techniques in Biophysics and Cell Biology* (R. Pain and B. J. Smith, eds). J. Wiley, London, Vol. 2, p. 1.

CHAPTER 3

LIGHT MICROSCOPY

G. B. Dealtry

1. GENERAL INTRODUCTION

The complexity and diversity of the optical microscopes and light microscopy procedures now available precludes the inclusion in this chapter of in-depth descriptions of all techniques, or of the detailed working of the microscope. For such information, the reader should refer to the excellent series of microscopy handbooks produced by the Royal Microscopical Society in conjunction with Oxford University Press (1–6) and to other reference texts (7–16). This chapter contains an overview and comparison of the major techniques, their application and limitations; with details of some of the more commonly used reagents.

2. BASIC TYPES OF MICROSCOPES

Two major types of light microscope have been developed that are optimized for specific functions (see *Table 1*). Developments of the inverted compound microscope include attachment of a heated stage and enclosed culture chamber (17), micro-manipulation systems (18), video and photographic image analysis and attachment of both inverted and upright compound microscopes to other analytical instruments (14, 15, 16).

Table 1. Basic microscope types

Stereomicroscope	Compound microscope
Two or more lens systems	Two or more lens systems
High-quality erect image	Inverted virtual image produced (a real image can be formed for projection and photography)
Binocular	Transmitted light usually used (reflected light is available)
Good depth of field	Contrast enhancement
Usually has a large-diameter common main objective lens	Good measurement facilities
Easy manipulation of objects (no lateral reversal of image)	Inverted forms available
Longest working distance	High numerical aperture
	Needs high-quality condenser, objective and eyepieces

2.1. Components of the microscope

Light source

Most basic microscopes use low-wattage, low-voltage bulbs. Light from the bulb is focused via a collector lens system onto the condenser.

Different light sources and their relative values are given in *Table 2.*

Table 2. Light sources

Tungsten filament	Tungsten–halogen	Mercury vapor	Xenon vapor
Cheap, easily available Emits a continuous spectrum from 300 nm to 1500 nm Greatest energy output in the red range Light output reduces with age as tungsten from the filament deposits onto the bulb	Single filament array in a quartz bulb Emits a continuous color spectrum Color temperature of 3200–3250 K suitable for color photography with tungsten light film High illumination, no blackening with age High running temperature needs ventillated lamp housing	Discharge tube contains pressurized mercury vapor Emits concentrated peaks of light at 365, 436 and 546 nm The 365 and 436 nm bands are used for fluorescence microscopy The 546 nm band produces monochromatic green light High illumination High running temperature	Discharge tube contains pressurized xenon vapor Emits a continuous color spectrum Color temperature approaches daylight (5300–6000 K) High illumination High running temperature

Condenser

In transmitted light microscopes the condenser concentrates an even cone of light of the same numerical aperture (NA, see below) as that of the objective onto the specimen. A range of condensers are available that provide combinations of differing levels of optical correction, power or NA, and adaptation for specialist use (see *Table 3*). In reflected light microscopes the objective serves as its own condenser.

Numerical aperture

The numerical aperture (NA) of an objective is a measure of the ability of the objective to collect light rays diffracted by passage through the specimen; the wider the angle of the rays collected, the better the resolution of the image. The NA is calculated as the product of the sin of α (half the acceptance angle of the objective lens) and the refractive index (RI) of the optical medium in which the objective lens is working:

$$NA = \sin \alpha \text{ (RI optical medium)}$$

Table 3. Condenser types

Optical correction	Power (NA max.)	Adaptation
Uncorrected	Low power (maximum NA of 0.25)	Bright-field
Achromatic correction	Medium power (maximum NA of 0.95)	Dark-field
Achromatic and aplanatic correction	High power (maximum NA of 1.4)	Phase-contrast
		Nomarski differential-interference contrast
		Hoffman modulation contrast

Since the RI of air is 1.0 (see *Table 4*), then the maximum NA for a dry objective lens with no cover slip over the specimen is 1.0 (sin α). However, use of a cover slip reduces the effective NA of the objective. Oil-immersion objective lenses can attain higher NAs (up to 1.4) by virtue of the fact that the RI of immersion oil is the same as that of glass and forms a homogeneous medium for the light rays to pass from the specimen to the lens.

Objective

Resolution. For most compound light microscopes the maximum lateral resolution is about 0.25 μm in the object plane, and the maximum vertical resolution about 5 nm if interferometric methods are used. The highest quality images are produced by objectives of high numerical aperture. Generally, magnification should be no greater than 1000 times the numerical apperture of the system. Since the maximum NA of a dry lens is 0.95 and that of oil-immersion lenses is 1.4, the magnification of most systems ranges from about $\times 10$ to $\times 1500$.

Types of objective. The main types of basic objective are compared in *Table 5*. Correction of the curvature of the field can be provided by plan-achromat, plan-fluorite and plan-apochromat objectives. Other objectives to provide corrections for different cover-slip

Table 4. Refractive indexes of immersion media[a]

Optical medium	Refractive index
Air	1.00
Glass (slide and cover slip)	1.51
Ethyl alcohol	1.33
Water	1.336
Xylol	1.49–1.5
Cedar oil	1.5–1.51
Immersion oil	1.515–1.52
Methylene iodide	1.74

[a] Data from ref. 19.

Table 5. Objectives

Achromat	Fluorite or semi-apochromat	Apochromat
Some correction for chromatic and spherical aberrations Cheap and easy to use Quite long working distance Best performance with green filters passing 550 nm wavelength light	Better correction for chromatic and spherical aberrations Shorter working distance High NA Have chromatic difference of magnification in the primary image, so need compensating eyepieces	Highest quality High NA Complete correction for color and spherical aberrations Also have chromatic differences of magnification

widths (or no cover slip), fluorescence-free objectives for reflected light fluorescence, and long working distance objectives are available.

Special objectives for phase-contrast microscopy, the use of polarized light, modulation contrast, interference contrast and Nomarski differential-interference contrast are also produced. In these cases appropriate matched condensers are also required (see Sections 3.1 and 3.2). The characteristics of typical objective lenses are given in *Table 6*.

Immersion objectives, even of relatively low magnification, are increasingly used for fluorescence microscopy because they have high NAs, giving brighter images:

$$\text{Brightness} \propto \frac{\text{NA}^2}{\text{Magnification}^2}$$

Magnification changers, although not part of the objective, add auxillary lenses to the tube to increase the magnification of the primary image.

Eyepiece
The eyepiece has a magnification range of ×8 to ×20. Many eyepieces also further correct aberrations remaining in the primary image produced by the objective. There are two main types of eyepiece; the Huyghenian (internal diaphragm) eyepiece and the Ramsden (external diaphragm) eyepiece. These are compared in *Table 7*.

3. LIGHT MICROSCOPE TECHNIQUES

3.1. Phase-contrast microscopy

Unstained live cells or thin tissue sections do not significantly alter the intensity (amplitude) or the color (wavelength) of light passing through them and so are poorly resolved by bright field illumination. Phase-contrast microscopy makes use of variations in the refractive index and thickness of the specimen that affect the relative phase of the light waves passing through it. It is particularly good for observation of living cells in culture. One drawback of the phase-contrast image is the formation of a bright halo at the edge of highly refractile components.

Table 6. Characteristics of different types of general-purpose objectives[a]

Objective type	Approx. focal length (mm)	Approx. numerical aperture	Approx. initial magnification
Achromat (dry)[b]	30.03	0.10	×4
	25	0.18	×6
	16.9	0.25	×10
	8.63	0.40	×20
	4.58	0.65	×40
	1.92	1.30	×100
Achromat (oil)[c]	2	1.25	×100
Plan-achromat (dry)[b]	36.54	0.13	×4
	18.98	0.30	×10
	8.03	0.46	×20
	4.13	0.70	×40
	1.68	0.95	×100
Plan-achromat (oil)[c]	3.8	0.90	×50
	1.75	1.25	×100
Fluorite (dry)[b]	4	0.75	×40
Fluorite (oil)[c]	2	1.30	×100
Plan-fluorite (dry)[b]	137.9	0.04	×1
	73.42	0.08	×2
Apochromat (oil)[c]	2	1.40	×90
Plan-apochromat (dry)[b]	36.71	0.16	×4
	16.92	0.40	×10
	7.68	0.70	×20
	4.18	0.95	×40
Plan-apochromat (oil)[c]	4	1.00	×40
	2.8	1.40	×60
	1.62	1.40	×100

[a] Data taken from ref. 7 and Olympus Optical Co. (UK) Ltd.
[b] Most objectives are used dry, i.e. with air between the front of the lens and the cover slip or specimen, and have a maximum theoretical NA of 1.0 (see earlier section). Most are designed for use with no cover slip or a cover slip of 0.17 mm thickness. High NA, dry apochromatic objectives have a correction collar for spherical aberrations caused by the use of cover slips of incorrect thickness, and/or variation in the thickness of the mountant.
[c] Objectives of high NA are designed for use with an immersion medium between the lens and the specimen. Immersion oil has the same refractive index as glass, and therefore avoids spherical aberration resultant from use of the incorrect cover-slip thickness (see earlier section). Other immersion media are listed in *Table 4*.

Table 7. Comparison of Huyghenian and Ramsden eyepieces

Huyghenian eyepiece	Ramsden eyepiece
Contains two planoconvex lenses and an internal field diaphragm	Contains two planoconvex lenses and an external field diaphragm
Both lenses have their convex surfaces directed towards the objective, with the diaphragm between the lenses	Convex surfaces of the lenses face each other with the diaphragm mounted below the lower lens
Provides a satisfactory image with achromatic and plan objectives	Allows the insertion of an eyepiece graticule without distorting the object image
Compensating versions are needed to correct the chromatic differences of magnification produced by fluorite and apochromatic objectives	Provides a satisfactory image with achromatic and plan objectives
	Compensating versions are needed to correct the chromatic differences of magnification produced by fluorite and apochromatic objectives

3.2. Interference microscopy

Interference microscopes overcome the formation of a halo of secondary interference found in phase-contrast images (see Section 3.1) by completely separating the direct and diffracted light rays with a beam-splitting device. By varying the relative phase and amplitude of the direct light rays, the degree of alteration of the diffracted rays can be measured and the refractive index and relative mass of a given cell component estimated. If the thickness of a cell sample is known, then the dry mass of a given component of the cell can be calculated from the formula:

$$M = \frac{\phi \times A}{100 \times \alpha}$$

where ϕ is the measured phase change; α, the specific refractive increment (a constant usually given the value 0.0018); A is the area over which ϕ was measured; and M is the dry mass of the component.

To determine the mass of an entire cell by this method, an automatic integrating instrument must be used to scan the entire cell and obtain an integrated value for the phase change over the whole cell.

3.3. Dark-ground microscopy

Dark-ground microscopy is also used to increase the contrast of unstained specimens. The specimen must show significant differences in its refractive index from that of the surrounding medium. The fine structure of aquatic organisms such as the protozoans *Stentor* and *Tetrahymena*, and the crustacean *Daphnia*, can be seen particularly clearly. A reversed contrast image is formed, with structures within and at the edges of the specimen showing brightly against a dark background. The image has a high degree of contrast and it is possible to detect fine structure that is not so easily detected by other forms of microscopy, although the true resolution is no different to that produced by the same instrument operated with bright-field illumination.

3.4. Polarizing microscopy

This technique is based on the ability of anisotropic components within a specimen to alter the phase of polarized light (formed by polarizing filters). Substances with a crystalline or laminated molecular structure, or containing a regular arrangement of submicroscopic rod-like structures, can differentially alter the velocities of two rays of polarized light vibrating in perpendicular planes. This produces birefringence, which normally appears white against a dark background; or, where the birefringent substance is of uniform thickness, interference colors are produced when the fast and slow rays are reunited.

Differences in light absorbance along different planes of an asymmetrical substance can be detected by dichroism under polarized light of a single plane. Either light of a specific wavelength is selectively absorbed, or there is a change in the intensity of white light when the light passes through the substance in certain planes.

3.5. Fluorescence microscopy

A large range of fluorescent probes, analogues, indicators and tracers are available for biochemical analyses and to localize directly or indirectly a vast array of molecules and cellular components.

Fluorescent compounds absorb short-wavelength light (usually in the blue to ultraviolet range of the spectrum) and transiently emit light of longer wavelength than the original excitation wavelength. Because fluorescence is intrinsically short-lived, samples must be shielded as much as possible from excitation wavelengths during preparation, stored for a limited time period prior to examination and viewed quickly with limited exposure to the excitation light.

Primary or autofluorescence

This is produced by a number of biological substances following exposure to blue or ultraviolet light (see *Table 8*). In mammalian cells most of the autofluorescence is due to NADH, riboflavin and flavin coenzymes which have excitation spectra ranging from 350 nm to 500 nm (20, 21). Proteins fluoresce when excited at 250 nm to 280 nm due to the presence of tryptophan, tyrosine and phenylalanine (22). Some use can be made of changes in autofluorescence to study molecular reactions (23) or to locate particular compounds, but these techniques are of limited value.

Table 8. Some autofluorescent substances[a]

Substance	Fluorescent emission color
Animal and plant tissue	Blue
Porphyrins	Strong red
Chlorophyll	Strong red
Collagen	Blue-green
Elastic fibres	Brilliant blue
Lipid droplets	Yellow
Ceroid	Shades of yellow
Lipofuscin	Orange
Various vitamins	Yellow, red, green or blue
Tetracycline	Bright yellow

[a] Data from refs 7 and 24.

Secondary fluorescence
This is either induced in a naturally non-fluorescent compound by chemical interaction, or formed by the application of an exogenous fluorescent probe to a specimen. Induced fluorescent reactions include the condensation of various amines with formaldehyde (7), for example to locate dopa, dopamine, catecholamine, adrenalin and 5-hydroxytryptamine. However, like primary fluorescence, this is a rather limited technique.

Fluorescent probes
A wide range of fluorescent probes is now commercially available (for suppliers, see Appendix). They are chosen for specific applications on the basis of their absorption and emission spectra, extinction coefficient (a measure of the probability of absorption and hence of emission), quantum yield (the efficiency of fluorescence), fluorescence lifetime, environmental effects, and chemical and biological activity. Some of the commoner fluorophores are listed in *Table 9.*

In addition, various fluorescent conjugates of antibodies, lectins, biotin, avidin and streptavidin, ferritin, albumin, dextrans, protein A, and a range of enzymes including peroxidase, alkaline phosphatase, β-galactosidase and enzyme substrates are available (see Appendix for major suppliers).

Fading of fluorescence is a common problem that can be reduced by mounting samples in anti-fade, composed of 2.5% (w/v) 1,4-diazabicyclo[2-2-2]octane (DABCO) in 90% glycerol, 10% phosphate buffered saline.

Table 9. Some commonly used fluorophores

Probe	Conditions[a]	Use	EX[b]	EM[c]	Ref.
Fluorescein-5-isothiocyanate (FITC isomer I)	pH 7, PBS	Low MW covalent amine reactive label	494	520	25 26 27
5-Iodoacetamido-fluorescein (5-IAF)	pH 7, PBS	Low MW covalent sulfhydryl reactive label	490	520	25 26 27
Fluorescein-5-maleimide	pH 7, PBS	Low MW thiol reactive label	490	515	25 26 27
Tetramethyl-rhodamine isothiocyanate (TRITC)	pH 7, PBS	Low MW covalent amine reactive label	541	572	25 27
Rhodamine X isothiocyanate (XRITC)	pH 7, PBS	As for TRITC (better resolution)	578	604	26 27
Texas red (sulfonyl chloride derivative of sulforhodamine 101)	pH 7, PBS	Covalent amine reactive label, forming stable sulfonamides	596	620	26 27 28

Table 9. Continued

Probe	Conditions[a]	Use	EX[b]	EM[c]	Ref.
Lissamine rhodamine B		Alternative to Texas red	567	584	27
Eosin-5-isothiocyanate	pH 8, 0.5 M NaCl	Covalent amine reactive label	524	548	27 29 30
NBD-amine	Ethanol, methanol	Covalent label	478	520–550	31 32 33
Coumarin-phalloidin	Water	Covalent label	387	470	34
Phycoerythrin-R	pH 7, PBS	Covalent label	480–565	578	27 35
Lucifer yellow CH lithium salt	Water	Tracer for cell coupling	428	540	27
Cascade Blue	Water	Polar tracer	375, 398	424	27
Hoechst 33342	Excess DNA	Nuclear stain binds DNA minor groove (AT selection)	340	450	26 27
4′,6-Diamidino-2-phenylindole, hydrochloride (DAPI)	Excess DNA	Nuclear stain	350	470	26 27
Ethidium bromide	Excess DNA	Nuclear stain (dead cells)	510	595	27 36
Propidium iodide	Excess DNA	Nuclear stain (dead cells)	536	623	26 27
Acridine orange	+ DNA + RNA	Stains live cells	480 440–470	520 650	37 38
Pyronin Y	+ dsDNA + dsRNA + ssRNA		549–561 560–562 497	567–574 565–574 563	39 40
Thiazole orange	RNA		453	480	26 41
1,1′-Dihexadecyl oxacarbocyanine, perchlorate (diO-Cn-(3))	Methanol	Cationic lipophilic membrane potential probe	485	505	26 42
Dil-Cn-(3)	Methanol	As above	548	567	26 42
Dil-Cn-(5)	Methanol	As above	646	668	26 41

Table 9. Continued

Probe	Conditions[a]	Use	EX[b]	EM[c]	Ref.
Rhodamine 123	Ethanol	Membrane fusion	556	577	26 43
Nile red	Heptane acetone	Lipid content fluidity	485 530	525 605	44 45
1,6-Diphenyl-1,3,5-hexatriene	Hexane	Neutral membrane probe	348	426	26 27
NBD phosphatidyl ethanolamine		Phospholipid	450	530	46
6-Carboxy-fluorescein	High pH Low pH	Intracellular pH	495 450	520 —	26 27
BCECF	High pH Low pH	As above	505 460	530 —	26 27
Seminaphtho-rhodafluor-1 (SNARF)-1	pH 5.5 pH 10	Color pH indicator	518–548 574	587 636	26 27
Fura-2	Low calcium High calcium	Intracellular calcium probe	335 360	512–518 505–510	27 47
Indo-1	Low calcium High calcium	As above	350 330	482–485 390–410	27 47
Fluo-3	Low calcium High calcium	Calcium indicator, fluorescent on hydrolysis	506 506	526 526	27

[a] Abbreviations: PBS, phosphate buffered saline; dsDNA, double-stranded DNA; dsRNA, double-stranded RNA; ssRNA, single-stranded RNA.
[b] EX = absorption maximum (nm).
[c] EM = emission maximum (nm).

Fluorescence microscopes

The design and use of fluorescent microscopes are well described in ref. 22. The microscope is a conventional compound microscope with at least two light sources, one to provide normal illumination and one for excitation of fluorophores. Additional fluorescence excitation light sources may be used for multiple spectral imaging (26). Illumination for fluorescence is usually by tungsten, quartz–halogen, 50 or 100 W mercury arc, or 75 W xenon arc lamps. Laser light sources are also used, with particular success in confocal microscopy (see Section 3.6).

Both transmitted light and incident light systems are available. Most commonly, incident light epi-illumination (epifluorescence) is used in cell biology. This has two advantages over transmitted light. First, the excitation wavelengths illuminate only the area of the specimen under observation, so other areas of the specimen are not photo-bleached. Secondly, because the exciting wavelengths are directed onto the surface of the specimen, there is no loss of fluorescence by absorption within the thickness of the specimen.

Excitation and emission filters may be neutral density, colored glass, gelatin or interference filters; and are classified by their transmission and reflection characteristics. The highest quality filters are the interference filters, which provide sharp cut-offs between the range of light wavelengths transmitted and those absorbed or reflected. Filters are chosen to isolate the excitation and emission spectra of the fluorophore in use. Where two or more fluorophores are to be visualized together, narrow band-pass filters may be needed. Examples of filter sets appropriate for some common fluorophores are given in *Table 10.*

Fluorescence may be detected visually or via low light level video-cameras (48) or slow-scan charge-coupled-device (CCD) cameras (49, 50), and processed by image analysis.

3.6. Confocal microscopy

Confocal scanning microscopy enables non-invasive optical sections with a depth of field of about 0.5 μm to be taken through a specimen. This avoids the blurring caused by inclusion in the image of out-of-focus regions. The specimen is scanned by light focused at a point at the focal plane within the specimen. Successive interactions between the light and the specimen are accumulated and integrated to form a confocal image within that plane. Serial optical sections can be used to form a three-dimensional construct. For detailed discussion of the principles and practice of confocal microscopy, see refs 16 and 51. This and other areas of light microscopy increasingly rely on the capture, manipulation, storage and analysis of images by computerized photodetection systems. The principles of these processes are well explained by Kennedy (52) and in greater detail in ref. 15.

Table 10. Filter sets for selected fluorophores[a]

Fluorescent probe	Excitation filters[b]	Dichroic mirror[c]	Emission filters[b]
Hoechst 33342	G350/56	395	420/LP
Fluorescein–actin	485/20	510	542/45
Lissamine–rhodamine–dextran	546/12	580	590/40
DilC$_1$ (5)	615/30	650	695/70
Cy7-dextran	720/40	759	785/50

[a] Data abstracted from ref. 26, giving filter sets used for five fluorescence parameters imaging.
[b] First number is the center wavelength of the bandpass filter; second number is the total bandpass at half maximum transmittance. LP is a long-pass glass filter, in which case the number is the wavelength of half maximum transmittance.
[c] Wavelength of half maximal transmittance of long-pass dichroic mirrors.

4. SAMPLE PROCESSING FOR LIGHT MICROSCOPY

4.1. Fixation

Living material can be examined sucessfully under the light microscope without fixation, for example by phase-contrast microscopy of unstained material or by using vital stains (see Section 3.1). But for more detailed examination and long-term storage of samples, material is usually fixed and stained or otherwise processed.

Fixation procedures are all directed towards preserving cells from deterioration, stabilizing cell structure, preventing loss of cell contents, permeabilizing the cell membrane and exposing reactive sites for detection with stains or other probes, while maintaining the specimen as near to its living state as possible.

Cell structure and content can be maintained largely by cross-linking or coagulating the cellular proteins. Combination of different fixatives which swell protein complexes (e.g. acetic acid) with those that cause shrinkage (e.g. methanol) is effective. Most fixatives also preserve nucleic acids and macromolecular carbohydrates. Lipids are readily dissolved and care is needed if these are to be retained; generally, frozen samples are used.

Most fixation techniques alter the size and shape of cells and cellular components significantly (*Table 11*). However, if processing is performed consistently, the changes induced will be common within cell types.

Fixatives
Specimens may be physically or chemically fixed.

Physical fixation methods.
(i) The simplest physical method is to heat the specimen over the flame of a Bunsen burner, in a microwave oven (53), or by immersion in boiling saline (13). This coagulates proteins and melts lipids, so does not maintain the cell structure well. However, it is a quick, easy method for fixing bacteria and is commonly used in diagnostic work.
(ii) The other simple method is air-drying. For good results cells should be dried quickly, so this technique is best suited to use with thin cell smears. The drying process is speeded up by using a hair-drier or fan heater. Cells should be dried in a physiological solution such as serum, or a solution containing 20% albumin to prevent destruction of nuclei in the hypertonic environment otherwise formed as the specimen dries. Addition of 80% methanol to the solution helps to preserve nuclei intact. Post-fixation with an alcohol such as methanol also improves the preservation of chromatin structure.

Table 11. Effect of fixation and staining on cell area[a]

Procedure	Nuclear area (μm^2)	Cell area (μm^2)	N/C ratio
Air-dried May–Grunwald–Giemsa	95 ± 11	27 ± 65	0.37 ± 0.07
50% Ethyl alcohol, Papanicolaou	49 ± 9	176 ± 46	0.30 ± 0.07
96% Ethyl alcohol, Papanicolaou	32 ± 6	114 ± 56	0.29 ± 0.06
Leiden spray fixative, Papanicolaou	40 ± 9	131 ± 42	0.031 ± 0.07
No fixation, no staining, phase contrast	60 ± 12	208 ± 72	0.34 ± 0.06
Methanol–acetic acid	78 ± 8	—	—

[a] Data from ref. 12.

(iii) Freezing and freeze-drying are also effective. Specimens must be frozen rapidly to minimize the formation of ice crystals, which can rupture membranes, then sectioned in a cryostat and processed. Procedures vary, generally small, fresh samples are treated by one of the following:

(a) direct immersion into liquid nitrogen (− 196°C);
(b) immersion in isopentane pre-cooled to its freezing point (− 170°C) in liquid nitrogen;
(c) placing on a support pre-cooled with liquid nitrogen or solid carbon dioxide (− 75°C);
(d) exposure to carbon dioxide gas; or
(e) treatment with aerosol spray containing difluorodiphenyl.

Addition of cryoprotectants such as dimethylsulfoxide (DMSO), glycerol and sucrose at 10% concentration (54) helps to reduce damage due to ice crystals. Specimens are then cryostat-sectioned.

(iv) The frozen specimen can be dehydrated by freeze substitution in ethanol, acetone or *n*-butanol below −49°C (13, 55). The ice is dissolved at this low temperature, but the proteins are unaffected. After complete dehydration the sample temperature is raised to 4°C and left for a few hours to allow chemical fixation to occur. Freeze substitution with *n*-butanol is preferred for some enzyme histochemistry techniques (56).

(v) Specimens can also be freeze-dried in a freeze-drying chamber where they are held in a vacuum at −40°C until the ice sublimes. Proteins are not insolubilized by this procedure, but water-soluble low molecular weight substances are retained. Subsequent fixation with a gas such as formaldehyde enhances specimen preservation.

Chemical fixation. The rate of penetration of a chemical fixative determines the size of specimen that can be handled. Generally, fixation should not take less than 15 min or exceed 24 h duration, except in the case of formaldehyde (12, 13). Rapidly penetrating fixatives can fix specimens of 3–5 mm thickness in 24 h; slower-penetrating fixatives should be used with specimens less than 2 mm thick (13).

Penetration rates of some common fixatives (in order of fastest to slowest (12)):

acetic acid > formaldehyde > mercuric chloride > osmium tetroxide > ethanol.

(i) Thin specimens such as cryostat sections, films and smears are immersed in the appropriate fixative (generally formaldehyde, ice-cold acetone, ethanol or methanol) either before or after staining.

(ii) Small samples can be fixed for 2–24 h in vapor above an aqueous solution of osmium tetroxide (13).

(iii) Phase-partition fixation uses a fixative dissolved in a liquid immiscible with water. The fixative passes from the non-aqueous phase into the aqueous phase of the specimen (for example, intracellular and extracellular fluids), reducing the loss of small proteins and amino acids and avoiding disrupting the ionic compositions of fluids within the specimen (57).

(iv) Larger specimens are immersed in at least twenty times their own volume of fixative.

(v) Animal samples can also be fixed by perfusion of the fixative through the vascular system (13).

Classification of chemical fixatives. These can be coagulant, cross-linking, additive (linking to the substance to be fixed) or nonadditive. The choice of fixative depends on the sample and staining technique to be used. Some of the major fixatives and their actions are listed in *Table 12.*

Table 12. Properties of major fixatives[a]

Property	Ethanol, methanol, acetone	Acetic acid	Formaldehyde
Normal concentration (% alone or in mix)	70–100	5–35	2–4
Penetration	Fast	Fast	Fairly rapid
Effect on tissue volume (relative)	Large decrease	Large increase	Nil
Hardening of sample (relative)	Great	Nil	Medium
Effect on proteins	Non-additive; coagulant; denatures most; gelatinizes histones; charge little changed	Nil	Additive; non-coagulant; cross-links via NH_2 and SH groups
Effect on nucleic acids	Denatures; solvation of bases increased, of riboses decreased	Precipitates	Slight extraction
Effect on carbohydrates	Nil	Nil	Nil
Effect on lipids	Some extraction; dilute ethanol splits lipids from lipoproteins	Nil	Preserves well
Effect on enzyme activities	Some preserved, if cold	Possible inhibition	Some preserved if short, cold exposure
Effect on organelles	Destroyed	Destroyed	Preserved
Use with anionic dyes	Satisfactory	Poor	Rather poor
Use with cationic dyes	Satisfactory	Good	Good
Use as sole fixative	Small blocks or cryostat section	Nil	Useful

[a] Data from refs 12 and 13.

Glutaraldehyde	Mercuric chloride	Osmium tetroxide
0.25–4	3–6	0.5–2
Slow	Fairly rapid	Slow
Nil	Slight decrease	Nil
Medium	Medium	Little
Additive; non-coagulant; gelatinizes histones; renders charge less negative	Additive; coagulant; nucleoproteins flocculate; cross-links; reacts with ionizable bases, with SH group	Additive; non-coagulant; cross-links via NH_2 and SH groups; renders charge less positive
Slight extraction	Coagulation through N of heterocycles; precipitates DNA weakly; nuclear mercuration possible	Slight extraction; can react with C=C groups of bases forming diols; DNA not precipitated
Possibly nil	Nil	Possibly some oxidation
Preserves well	Does not insolubilize; hydrolyses plasminogen; useful with Schiff's reagent	Addition to and oxidation of C=C bonds giving black color
Most inhibited	Inhibition	Inhibition
Well preserved	Preserved	Well preserved
Rather poor	Good	Acidophilia changed to basophilia
Satisfactory	Good	Satisfactory
Rarely used alone	Poor	Specialized uses only

4.2. Processing

Light microscopy depends upon two-dimensional examination of specimens, therefore samples must either be spread onto the microscope slide or sectioned and applied (for a comparison of these approaches see ref. 12). Some approaches to the preparation of biological specimens for microscopy are shown in *Table 13*. Retention of cells on the microscope

Table 13. Preparation of biological specimens[a]

Specimen	Procedure
Hard solids	
Mineralized bone	Thick section, attach to slide, grind down
Woody plant stem	Soften, cut on a microtome or macerate and suspend to give a thin section or cell suspension
Solids	
Cervical epithelium	Take a swab of superficial cells, smear onto slide
Solid organ (e.g. liver or kidney)	Either freeze and section on a cryostat, or embed and section on a microtome
Soft solids	
Bone marrow	Disperse cells in buffer to a single-cell suspension, smear onto a slide
Spinal cord	Wipe across a slide to deposit a cell monolayer
Fluids	
Cell-rich (e.g. blood)	Smear or spin onto a slide to give a cell monolayer
Cell-poor (e.g. urine)	Pass through a filter and transfer cells from the filter onto an albumin-coated slide
To stabilize	
Glycosaminoglycans	Precipitate with cetyl pyridinium salts
Lipids	Avoid organic solvents; section on cryostat; or cross-link with osmium tetroxide or potassium dichromate fixatives
Nucleic acids	Trap by fixing associated proteins; fix with weakly acid fixatives (e.g. acetic acid–methanol)
Non-ionic polysaccharides (e.g. glycogen)	Trap by fixing surrounding proteins; use non-aqueous solutions (e.g. formal alcohol)
Proteins	Use denaturing and/or cross-linking fixatives (e.g. alcohol or glutaraldehyde)
Protein antigens	Check antigen/antibody reactivity with selected fixatives, if in doubt use unfixed cryostat sections or cell preparations; fix in formaldehyde and expose antigen with a protease
Catalytically active enzymes	Use unfixed cryostat sections or cell preparations; use a colloid protecting agent on semi-permeable membranes

[a] Data from ref. 58.

Table 14. Cell loss during slide preparation (urothelial carcinoma)[a]

Preparation	Percentage of original cell number
Wet fixation, albumin-coated slides, Papanicolaou stain	4–13
Wet fixation, Papanicolaou stain	4–18
Wet fixation, 96% alcohol, Papanicolaou stain	2–44
Spray fixation, Papanicolaou stain	16–60
Air drying, Papanicolaou stain	43–100
Air drying, albumin-coated slides, May–Grunwald–Giemsa stain	58–99
Air drying, May–Grunwald–Giemsa stain	76–96
Millipore filter, Papanicolaou stain	65–95

[a] Data from ref. 12.

slide is a problem with cell spreads and smears (for an indication of cell loss during preparation see *Table 14*).

Initial adherence of cells to the slide (12) can be increased by:

(i) coating the slide with poly-L-lysine, ovalbumin, or gelatin–chrome alum;
(ii) centrifuging the specimen onto the slide (in a Leif centrifuge bucket or a Cytofuge);
(iii) using cold slides from the freezer;
(iv) using polyethylene glycol in a spray fixative;
(v) using unfixed cells;
(vi) retaining mucus or proteins associated with any cell suspensions or smears;
(vii) air drying.

Cells cultured on cover slips attach more successfully if the cover slips are pre-coated with gelatin.

Typically, tissue sections are cut on a microtome or cyostat at a thickness of 4–6 μm. Prior to sectioning, samples are either frozen (see Section 4.1.) and processed by cryostat sectioning; or are embedded in gelatin, agar, OCT, Carbowax, polyacrylaminde, or paraffin wax (10, 13, 59). Dehydration of the specimen is not necessary with the water-soluble support materials (the first four materials listed), but is essential to allow paraffin to enter the specimen. Generally, ethanol of various concentrations, up to 100%, is used. Other dehydrating agents include dioxane, cellosolve, acetone and isopropanol (10, 13, 59). Clearing agents miscible with both the dehydrating agent and paraffin are used where the dehydrating agent (e.g. ethanol) is not itself miscible with paraffin (10, 13, 59). For embedding, sectioning and mounting procedures refer to Sheehan and Hrapchak (10), Kiernan (13) and Bancroft and Stevens (11). Prior to staining, the removal of some fixatives or by-products of fixation is necessary, as detailed by Kiernan (13).

4.3. Staining

Staining involves the selective uptake and interaction of reagents from a soluble phase with a solid specimen (e.g. the attachment of a dye or the binding of an antibody), ultimately to produce a specific visible change to the specimen which enables the identification of particular components. The more complex interactions involved in enzyme histochemical techniques and immunochemical methods are discussed in Sections 4.6 and 4.7.

Dye molecules contain a chromogen, the colored component, and an auxochrome, which attaches the dye to the specimen. The auxochrome either forms covalent or coordinate bonds with the specimen, or it is an ionizable component. The chromogen contains a chromophore that is involved in the absorption of certain wavelengths of light to give the color (see refs 12 and 13 for more theory). Dyes have been divided into acidic and basic groups, although, since dyes are usually salts, they can be more accurately described as anionic and cationic. A dye is basic (or cationic) when the chromophore is the cationic component. Some chromophores are amphoteric, in which case the dye reacts as an acid if the pH of the solvent is above the dye's isoelectric point, and a base if the pH is below its isoelectric point. Some commonly used dyes are listed in *Table 15*, with the classification of their chromophore.

Progressive staining
The rate of diffusion of the dye into the specimen determines the staining of a given component. Short exposure results in staining of only fast-staining components of the

► p. 45

Table 15. Common dyes[a]

Chromophore	Dye[b]	Acid/base/ amphoteric[c]	Use
Quinoid ring			
Triaryl-methanes	Basic fuchsin (i 288; 545 nm; w)	Basic	Schiff's reagent and nuclear
	Pararosanilin (i 288; 545 nm; w, a)	Basic	Nuclear
	Crystal violet (i 372; 589–593 nm; w, a)	Basic	Nuclear, vital and Gram's stain
	Methyl green (i 387; 420 and 630–634 nm; w)	Basic	Differential stain for DNA (with Pyronin Y for RNA)
	Solochrome cyanin R (i 467; red; w)	Acid	Nuclear and cytoplasmic
	Acid fuchsin (i 540; 540–546 nm; w, a, c)	Acid	Cytoplasmic
	Coomassie brilliant blue (m 818; 590 nm; w, a)	Acid	Cytoplasmic and quantitative protein stain
	Aniline blue (i 692; 600 nm; w)	am	Cytoplasmic
	Fast green (i 763; 625 nm; w, a)	am	Cytoplasmic
Hematein	Hematein (i 300; 445 nm; w, a, g, c)	am	Nuclear hematoxylin dye component

Table 15. Continued

Chromophore	**Dye**[b]	**Acid/base amphoteric**[c]	**Use**
Anthraquinones	Alcian blue (i 1380; blue; w)	Basic	Mucopolysaccharides
	Luxol fast blue (—; —; a, c, m, ac, ip)	Basic	Fats (polar lipids)
	Cuprolinic blue (i 640; 635 nm, w)	Basic	Nucleoli, with $MgCl_2$ is specific for RNA
	Alizarin blue (i 291 without $2NaHSO_3$; blue; w)	Acid	Nuclear mordant (Fe) dye metachromatic stain
	Purpurine	Acid	Demonstration of calcium
Xanthenes	Pyronin Y (i 268; 552 nm; w, a)	Basic	RNA
	Acridine orange (i 266; 467–497 nm; w, a)	Basic	Nucleoli fluorochrome
	Eosin Y (i 646; 515 nm; w, a)	Acid	Cytoplasmic counterstain
	Gallein (i 361; blue with Fe mordant; w (hot), a, ac, alkali)	Acid	Nuclear, glycogen
Azines and related dyes (quinone-imines)			
Azines	Neutral red (—; 530 nm; w, a)	Basic	Vital stain
Oxazines	Rhodaniline blue (m 780; —; w)	Basic	Nuclear and cytoplasmic
	Nile blue (i 318; 640 nm; w (hot), a)	Basic	Fat, sulfate, differentiates neutral fats and fatty acids
	Cresyl violet (m 321; 600 nm; w, a)	Basic	Metachromatic vital stain; at graded pH, protein stain
	Brilliant cresyl blue (m 332; 630 nm; w, a)	Basic	Vital stain; stains acid mucopolysaccharides
Thiazines	Azure A (i 256; 630 nm as orthochromatic dye; w, a)	Basic	Nuclear; methylene blue derivatives; RNA stain with a surfactant

Table 15. Continued

Chromophore	Dye[b]	Acid/base amphoteric[c]	Use
	Azure B (1) (i 260; 650 nm; w, a)	Basic	Nuclear; original component of Giemsa
	Methylene blue (i 284; 660 nm; w, a, c, g)	Basic	Nuclear and vital stain
	Toluidine blue (i 270; 620 nm; w, a, c, g)	Basic	Nuclear; RNA stain with a surfactant
Azo groups			
Azines	Azocarmine	Acid	Nuclear dye in Heidenhain's Azan
Monoazo-	Janus green B (i 300; 400 nm and 610–623 nm; w, a)	Basic	Vital stain
	Ponceau de xylidine (i 434; 500 nm; w)	Acid	Cytoplasmic
Diazo-	Bismarck brown Y (i 349; —; w, a)	Basic	Cytoplasmic vital stain
	Sudan black B (i 456; 600 nm; a, pg)	Basic	Fat stain
	Amido black (i 571; blue; w, a, c)	Acid	Nuclear (nucleoli); protein stain
	Oil red O (i 409; 550 nm; a, ac, ip)	Acid	Fat stain
	Biebrich scarlet (i 510; 503 nm; w)	Acid	Cytoplasmic; stains basic proteins at graded pH; pH; stains Barr bodies
Triazo-	Chlorazol black E (i 735; black; w, a, c)	Acid	Nuclear and cytoplasmic nitro-group
	Picric acid (i 228; 360 nm in alcohol; w, a, b, ch, e)	Acid	Fixative in Bouin's fixative and cytoplasmic
	Naphthol yellow S (i 312; 420 nm; w, a)	Acid	Protein, can be used with Feulgen reaction
	Dinitrofluorobenzene (i 186; 400 nm; w, a)	Acid	Protein

[a] Data derived from ref. 12.
[b] Data in brackets: ionic (i) or molecular (m) weight; absorption maximum (nm); soluble in water (w), alcohol (a), cellosolve (c), acetone (ac), methanol (m), glycol (g), isopropanol (ip), benzene (b), propylene glycol (pg), chloroform (ch), or ether (e).
[c] Amphoteric (am).

specimen. Best results with this approach are obtained with large or aggregated dyes that diffuse slowly. Accurate standardized timing is important for this. Combinations of dyes can be used.

Regressive staining
This involves long exposure to a dye beyond the staining equilibrium of all the components of the specimen, followed by partial destaining. Components with weak links to the dye destain first, revealing other, more strongly linked, components. High dye concentrations are used to give high diffusion rates into the specimen. Destains include water (better for acid than basic dyes), and alcohols (better for the basic dyes). The differentiation (destaining) times must be controlled carefully.

Differential staining
Two or more dyes are used to identify different components of a specimen (e.g. hematoxylin, nuclear; eosin, cytoplasmic). Details of staining methods are given in refs 10–13.

4.4. Microscopic tests for animal cell viability

Cell viability can be determined by observation of an untreated cell suspension (typically in a hemocytometer or cavity slide) under phase contrast; viable cells appear brightly refringent while dead cells are dark. Staining techniques involve either dye exclusion, in which the membrane of a viable cell excludes the dye which passively diffuses into dead cells, or specific uptake or passive infiltration of a vital stain.

All procedures should be performed rapidly (not longer than one hour before viewing) and use low-concentration (0.01%) solutions of dye in physiological buffer to minimize cell damage. Saline solutions of most dyes do not keep well, so stock solutions of 0.5–1% in water or other appropriate solvent should be diluted to between 0.01% and 0.2% in buffer immediately prior to use. Some common stains are described in *Table 16*. Cell suspensions, whole unicellular organisms and small tissue pieces (unless otherwise stated) can be examined.

4.5. Some common histological staining methods

(NB all the staining solutions in this section are available commercially, see Appendix for suppliers.)

Romanowsky–Giemsa (12)
Stock solution. Giemsa stain, store dark at room temperature. Lowering the pH to 4 with hydrochloric acid improves stability.
Working solution. Dilute stock 1 in 16 with Hepes buffer, pH 6.8

Use. General staining, nuclei stain purple, cytoplasm blue, although some specific variations occur, for example mucin stains reddish-blue.

Blood and bone-marrow stain, identifies cell types (12). General samples are fixed in methanol for 8 min, stained in stock solution diluted with an equal volume of buffer (50 mM NaH_2PO_4, 50 mM Na_2HPO_4, pH 6.8) for a few minutes, rinsed and air dried.

Blood and bone-marrow samples are post-fixed in stock solution for 5 min, rinsed in distilled water and stained in working solution for 25–35 min before rinsing and air drying.

Table 16. Staining tests for animal cell viability

Stain	Action	Ref.
Methylene blue 1.5 g methylene blue, 30 ml 95% alcohol, 2 ml 0.1 M potassium hydroxide; dissolve dye in alcohol, add potassium hydroxide, store in dark at 4°C	Vital stain	60, 62
Toluidine blue 0.5 g toluidine blue, 20 ml 95% alcohol, 80 ml distilled water; dissolve dye in alcohol, add water, filter and store in dark at 4°C	Vital stain	60, 61
Janus green 0.25% (w/v) stock aqueous solution	Vital stain Mitochondria stain green	62
Trypan blue Stock 2% (w/v) solution in distilled water, filter, add 1 volume solution to 9 volumes of cell suspension, incubate for 30 min at 37°C	Dye exclusion	63
Nigrosin 2% stock solution in phosphate buffered saline, dilute to 0.2% for use	Dye exclusion	64
Ethidium bromide/acridine orange 0.01 mg ml^{-1} ethidium bromide, 0.005 mg ml^{-1} acridine orange, cell suspension in holding medium containing 40% serum; mix 25 μl cell suspension with 10 μl ethidium bromide, after 30 sec add 10 μl acridine orange, mix and view	Differential fluorescent stain under ultraviolet light; dead cells red, live cells green	65

Papanicolaou's staining method (60)
Stock solutions. Harris's hematoxylin, store at room temperature; 0.1% hydrochloric acid; saturated solution of lithium carbonate; Orange G, store at room temperature; eosin azure-65 (EA-65).

Use. General nuclear and cytoplasmic stain. Nuclei are stained blue, nucleoli blue or red, and cytoplasm pink or blue, depending on cell type.

Slide preparations of samples fixed in 96% alcohol are hydrated through 70% and 50% alcohol to distilled water, then stained in hematoxylin for 6 min. After rinsing in running tap-water, then distilled water, slides are left in distilled water for 10 min. Staining is differentiated in 0.1% HCl until the color becomes red, then the slides are rinsed in distilled water, treated with lithium carbonate for 2 min, and dehydrated through the reverse series of alcohol concentrations to that used to hydrate.

Samples are then stained in Orange G for 2 min, dipped 15 times in 96% alcohol, and stained in EA-65 for 4 min, followed by treatment in two changes of 96% alcohol. Slides are left in absolute alcohol for 2 min, cleared in xylene and mounted.

Delafield's hematoxylin and eosin (12)
Stock solutions. Delafield's hematoxylin, dilute with an equal volume of distilled water before use; acid alcohol (three drops of concentrated HCl in 50 ml 95% alcohol); 0.1–0.5% eosin Y in 25% alcohol.

Use. General nuclear and cytoplasmic stain. Nuclei are stained dark blue and cytoplasm pink-red.

Fix slide preparations in 96% alcohol, wash in water, then stain in hematoxylin for 15 min. Rinse in tap-water, then stand in fresh tap-water for 10 min. If samples are still blue, add 2 drops of acid alcohol. Stain in eosin Y briefly, rinse in distilled water, dehydrate, clear in xylene and mount.

The Feulgen Reaction (12)
Stock solutions. Schiff's reagent, store at 5°C, discard if the solution turns pink; 1 M HCl at 60°C; 0.05 M metabisulfite; 0.01% Fast Green FCF in 95% alcohol.

Use. Locates DNA; DNA-rich sites are red, basic cell components are green.

Samples may be fixed in any fixative. Hydrolyze in HCl for 10 min, then leave in Schiff's reagent for 10 min. After washing in three changes of metabisulphite for 2 min each, samples are rinsed in running water, counterstained for a few seconds in Fast Green, dehydrated, cleared and mounted.

Methyl, green–Pyronin Y (66)
Stock solutions. Methyl green–pyronin Y solution, stored in a dark stoppered bottle; tertiary butanol–alcohol 3:1.

Use. DNA and RNA are differentially stained. Chromatin stains blue-green and nucleoli pink.

Samples should be fixed in Carnoy's fixative, then brought to water and stained in methyl green–pyronin Y for 3–5 min. Rinse slides in distilled water, blot and differentiate in the butanol–alcohol mixture for at least 2 min. Clear in two changes of xylene (2 min each), and mount in synthetic resin.

Periodic acid–Schiff reaction (12)
Stock solutions. Schiff's reagent (see above); 1% periodic acid in water; 0.52% $NaHSO_3$ (0.05 M); Mayer's hematoxylin.

Use. Stains glycogen-rich sites red, nuclei blue or black, mucin red-purple and cytoplasm gray/blue.

Samples may be fixed by any method, if air-dried they should be post-fixed in 70% alcohol for 10 min. All preparations are rinsed in water prior to oxidation in periodic acid for 5 min. Samples are then rinsed in distilled water, placed in Schiff's reagent for 15 min, put directly into three successive baths of $NaHSO_3$ for 2 min each, then rinsed in running tap-water for 10 min. Counterstain in hematoxylin, rinse in running water as before, dehydrate, clear in xylene and mount.

Sudan black B (67)
Stock solutions. Sudan black B (66); propylene glycol; glycerine jelly (40 g gelatin dissolved in 210 ml distilled water with 120 ml glycerine).

Use. Stains fat; neutral fats are green/brown, myelin, other lipids and mitochondria green/black, and cytoplasm unstained.

Formalin-fixed samples are washed in distilled water for 2–5 min, dehydrated in propylene glycol with agitation for 3–5 min, and stained in Sudan black solution for 5-7 min with agitation. The stain is differentiated in 85% propylene glycol for 2–3 min, rinsed in distilled water for 3–5 min, counterstained with nuclear stain if desired, and mounted in glycerine jelly.

Gram stain (60)
Stock solutions. Crystal violet; Lugol's iodine; 1% neutral red in water; acetone.

Use. To distinguish between Gram-positive and Gram-negative bacteria.

Air-dried, fixed slide preparations are equilibrated in water and stained in crystal violet for 1–2 min, and washed in tap-water (10 brief dips). Slides are then placed in Lugol's iodine for 0.5–2 min, washed in tap-water as before, dipped in acetone 10–15 times, and rinsed in running tap-water for 5–10 min, prior to staining in neutral red for 0.5–1 min. The preparations are finally washed in two changes of tap-water as before, dehydrated in two changes of 95% alcohol, two changes of absolute alcohol and two changes of xylene (dipping 15 times in each solution), and mounted.

4.6. Enzyme histochemistry

A large number of specialist enzyme detection systems are now developed for the cellular and subcellular localization of enzymes. The reader should consult appropriate texts for details.

General considerations include the need to use unfixed frozen specimens, or to choose a fixative that does not inhibit the enzyme (some phosphatases survive acetone fixation and most carboxylic ester hydrolases can be detected in formaldehyde-fixed preparations). Cold fixation retains enzyme activity most effectively. Freeze-drying followed by cold glycol methacrylate embedding also retains many enzyme activities (13, 68). Where the enzyme is detected by reactants in solution, a trapping agent is often included that reacts rapidly with the enzyme reaction product to deposit an insoluble colored product within the sample at the site of enzyme activity. Where the enzyme is detected by reaction with a substrate carried in a film applied to the specimen, the enzyme should be able to diffuse into the detection film, the film should be as thin as possible and the change produced must be immediate and irreversible. An adaptation of this is the electrophoretic transfer of ribonuclease activity from a cryostat section to an acrylamide gel containing RNA (69). Enzymes of homoiothermic animals generally react optimally at 37°C, whereas those of poikilotherms and plants function best at 25°C. But reduction of the temperature to 4°C does reduce diffusion of reactants. Samples must be kept moist during the reaction period.

4.7. Immunocytochemistry

This topic is covered by another book in this series (70), see also texts by Harlow and Lane (71) and Polak and van Noorden (72). Immunocytochemistry enables localization within a cell or tissue of any antigen to which an antibody is available. The antibody–antigen complex formed may be detected directly by attaching a marker to the reacting antibody or, more commonly, indirectly via a labeled secondary antibody raised against the immunoglobulin of the animal species providing the primary antibody. The latter technique is more sensitive and convenient, since a small panel of secondary antibodies to immunoglobulins of one species can be used to detect any primary antibodies in each of the different antibody classes. Single pan immunoglobulin antibodies that detect all immunoglobulin classes of a given species are also available, but these may be less sensitive than the class-specific antibodies. A large range of secondary antibodies labeled with fluorescent markers and enzyme markers are available (see Appendix for major suppliers). Fluorescent labels locate antigen more accurately than other labels and are ideal for confocal microscopy, but do not include the amplification steps of the other labels, or provide a permanent preparation, they are costly and require the use of a fluorescence microscope.

Fixation procedures should be mild to leave the antigen unaltered. However, permeabilization of the cell membrane by acetone, formaldehyde or methanol/acetone mix is necessary when detecting intracellular antigens, in order to allow entry of the antibody molecules. Antigens have differing fixative sensitivities, so a range of fixatives should be tested with each new antibody used.

5. TYPICAL DIMENSIONS OF CELLS AND ORGANELLES

The general sizes of procaryote and eucaryote cells and organelles are listed in *Table 17*. Exceptions to these sizes do exist, for example some bacteria can be up to several hundred microns in length (73) and mitochondria in filamentous fungi may be up to 30 μm long (74).

Table 17. General cell and organelle dimensions

Cell or organelle	Dimension (μm)	Ref.
Procaryote cell	0.15–5	75
Eucaryote cell	10–100	76
Nucleus	5–25	76
Chloroplast	2–8	76
Mitochondrion	1–10	76
Golgi apparatus	1	76
Lysosomes and peroxisomes	0.2–0.5	76
Plant cell wall	0.1–10	76

6. REFERENCES

1. Bradbury, S. (1989) *An Introduction to the Optical Microscope*. Oxford Science Publications, Oxford.

2. James, J. and Tas, J. (1984) *Histochemical Protein Staining Methods*. Oxford Science Publications, Oxford.

3. Bayliss High, O. (1984) *Lipid Histochemistry*. Oxford Science Publications, Oxford.

Light Microscopy

4. Ploem, J.S. and Tanke, H.J. (1987) *Introduction to Fluorescence Microscopy.* Oxford Science Publications, Oxford.

5. Polak, J. and van Noorden, S. (1987) *An Introduction to Immunocytochemistry: Current Techniques and Problems.* Oxford Science Publications, Oxford.

6. Thompson, D.J. and Bradbury, S. (1987) *An Introduction to Photomicrography.* Oxford Science Publications, Oxford.

7. Bradbury, S. (1976) *The Optical Microscope in Biology.* Edward Arnold, London

8. Lacey, A.J. (1989) *Light Microscopy in Biology: a Practical Approach.* IRL Press, Oxford.

9. Gurr, E. (1973) *Biological Staining Methods.* Searle Diagnostic, England.

10. Sheehan, D.C. and Hrapchak, B.B. (1987) *Theory and Practice of Histochemistry.* Battelle Press, Columbus.

11. Bancroft, J.D. and Stevens, A. eds (1977) *Theory and Practice of Histological Techniques.* Churchill Livingstone, Edinburgh.

12. Boon, M.E. and Drijver, J.S. (1986) *Routine Cytological Staining Techniques.* MacMillan Education, London.

13. Kiernan, J.A. (1990) *Histological and Histochemical Methods.* Pergamon Press, Oxford.

14. Wang, Y.-L. and Lansing Taylor, D. eds (1989) *Methods Cell Biol.*, **29**.

15. Wang, Y.-L. and Lansing Taylor, D. eds (1989) *Methods Cell Biol.*, **30**.

16. Pawley, J.B. ed (1990) *Handbook of Biological Confocal Microscopy.* Plenum Press, New York.

17. McKenna, N.M. and Wang, Y.-L. (1989) *Methods Cell Biol.*, **29**, 195.

18. Cellis, J.E., Graessmann, A. and Loyter, A. eds (1988) *Microinjection and Organelle Transplantation Techniques: Methods and Applications.* Academic Press, London.

19. Wilson, M.B. (1976) *The Science and Art of Basic Microscopy.* American Society for Medical Technology, Texas.

20. Benson, R., Meyer, R., Zaruba, M. and McKhann, G. (1979) *J. Histochem. Cytochem.*, **27**, 44.

21. Aubin, J. (1979) *J. Histochem. Cytochem.*, **27**, 36.

22. Lansing-Taylor, D. and Salmon, E.D. (1989) *Methods Cell Biol.*, **29**, 207.

23. Kohen, E., Thorell, B., Woiters, J., Kohen, C., Bartick, P., Salmon, J.-M., Viallet, P., Schatschabel, D., Rabinovitch, A., Minty, D., Meda, P., Westerhoff, H., Nestor, J. and Ploem, J. (1981) in *Modern Fluorescence Spectroscopy.* (E.L. Wehry, ed.). Plenum Press, New York, Vol. 3, p. 295.

24. Hrapchak, R. (1980) in *Theory and Practice of Histotechnology.* (D.C. Hrapchak and B.B. Sheehan eds). Battelle Press, Columbus, OH, p. 310.

25. Haugland, R.P. (1983) in *Excited States of Biopolymers.* (R.F. Steiner, ed.). Plenum, New York, p. 29.

26. Waggoner, A., DeBiasio, R., Conrad, P., Bright, G.R., Ernst, L., Ryan, K., Nederlof, M. and Taylor, D. (1989) *Methods Cell Biol.*, **30**, 449.

27. Haughland, R.P. (1989) *Handbook of Fluorescent Probes and Research Chemicals.* Molecular Probes, Oregon.

28. Titus, J.A., Haughland, R.P., Sharrow, S.O. and Segal, D.M. (1982) *J. Immunol. Methods*, **50**, 93.

29. Cherry, R.J., Cogoli, A., Oppliger, M., Schneider, G. and Semenza, G. (1976) *Biochemistry*, **15**, 3653.

30. Garland, P.B. and Moore, C.H. (1979) *Biochem J.*, **183**, 561.

31. Kenner, R.A. and Aboderin, A.A. (1971) *Biochemistry*, **10**, 4433.

32. Allen, G. and Lowe, G. (1973) *Biochem. J.*, **133**, 679.

33. Bratcher, S.C., Nitta, K. and Kronman, M.J. (1979) *Biochem. J.*, **183**, 255.

34. Small, J.V., Zobeley, S., Rinnerthalen, G. and Faulstich, H. (1988) *J.Cell Sci.*, **89**, 21.

35. Oi, V., Glazer, A.N. and Stryer, L. (1982) *J.Cell Biol.*, **93**, 981.

36. Pohl, F.M., Jovin, T.M., Baehr, W. and Holbrollk, J.J. (1972) *Proc. Natl. Acad. Sci.*, **69**, 805.

37. Kapuscinski, J., Darzynkiewicz, Z. and Melamed, M. (1982) *Cytometry*, **2**, 201.

38. Shapiro, H.M. (1985) *Practical Flow Cytometry.* Liss, New York.

39. Darzynkiewicz, Z., Kapuschinski, J., Tragonos, F. and Crissman, H.A. (1987) *Cytometry*, **8**, 138.

40. Kapuscinski, J. and Darzynkiewicz, Z. (1987) *Cytometry*, **8**, 129.

41. Lee, L.G., Chen, C.-H. and Chiu, L.A. (1986) *Cytometry*, **7**, 508.

42. Sims, P.J., Waggoner, A.S., Wang, C.H. and Hoffman, J.F. (1974) *Biochemistry*, **13**, 3315.

43. Kubin, R.F. and Fletcher, A.N. (1983) *J. Luminescence*, **27**, 455.

44. Greenspan, P. and Fowler, S.D. (1985) *J. Lipid Res.*, **26**, 81.

45. Sackett, D.L. and Wolff, J. (1987) *Anal. Biochem.*, **167**, 228.

46. Struck, D.K., Hoekstra, D. and Pagano, R.E. (1981) *Biochemistry*, **20**, 4093.

47. Grynkiewicz, G., Poenie, M. and Tsien, R.Y. (1985) *J. Biol. Chem.*, **260**, 3440.

48. Hiraoka, Y., Sedat, J.W. and Agerd, D.A. (1987) *Science*, **238**, 36.

49. Spring, K.R. and Lowy, R.J. (1989) *Methods Cell Biol.*, **29**, 270.

50. Aikens, R.S., Agard, D.A. and Sedat, J.W. (1989) *Methods Cell Biol.*, **29**, 292.

51. Shotton, D.M. (1989) *J. Cell Sci.*, **94**, 175.

52. Kennedy, J.M.T. (1991) *Image Enhancement Analysis*, **26**, 5.

53. Bernhard, G.R. (1974) *Stain Technol.*, **49**, 215.

54. Terracio, L. and Schwabe, K.G. (1981) *J. Histochem. Cytochem.*, **29**, 1021.

55. Bancroft, J.D. (1977) in *Theory and Practice of Histological Techniques* (J.D. Bancroft and A. Stevens eds). Churchill Livingstone, Edinburgh, p. 65.

56. Klaushofer, K. and von Mayersbach, H. (1979) *J. Histochem. Cytochem.*, **27**, 1582.

57. Nettleton, G.S. and McAuliffe, W.G. (1986) *J. Histochem. Cytochem.*, **34**, 795.

58. Horobin, R.W. (1989) in *Light Microscopy in Biology: A Practical Approach* (A.J. Lacey ed.). IRL Press, Oxford, p. 137.

59. Bradbury, P. and Gordon, K. (1977) in *Theory and Practice of Histological Techniques* (J.D. Bancroft and A. Stevens eds). Churchill Livingstone, Edinburgh, p. 29.

60. Koss, L.G. (1979) *Acta Cytologica*, **19**, 557.

61. Harris, M.J. and Keebler, C.M. (1976) in *Compendium on Cytopreparatory Techniques* (C.M. Keebler, J.W. Reagan and G.L. Wied eds). Tutorials on cytology, Chicago, p. 45.

62. Sumner, A.T. and Sumner, B.E.H. (1969) *A Laboratory Manual of Microtechnique and Histochemistry.* Blackwell, Oxford.

63. Kissmeyer-Nielsen, F. and Thorsby, E. (1970) *Transplantation Rev.*, **4**, 121.

64. Kaltenbach, J.P., Kaltenbach, M.N. and Lyons, W.B. (1958) *Exp. Cell Res.*, **15**, 112.

65. Weir, D.M. (1978) *Handbook of Experimental Immunology (3rd edn).* Blackwell, Oxford.

66. Clark, G. (1981) *Staining Procedures.* Williams and Wilkins, Baltimore.

67. Lillie, R.D. and Fulmer, H.M. (1976) *Histopathologic Technique and Practical Histochemistry.* McGraw Hill, New York.

68. Murray, G.I., Burke, M.D. and Ewen, S.W.B. (1986) *Histochem. J.*, **18**, 434.

69. Schultz-Harder, B. and Graf von Keyser-Lingk, D. (1988) *Histochem.*, **88**, 587.

70. Kerr, M. and Thorpe, R. (1992) *Immunochemistry Labfax.* BIOS Scientific, Oxford, in press.

71. Harlow, E. and Lane, D. (1988) *Antibodies: a Laboratory Manual.* Cold Spring Harbor Laboratory Press, New York.

72. Polak, J.M. and van Noorden, S. (1986) *Immunocytochemistry Modern Methods and Applications.* Wright, Bristol.

73. Linton, A.H., Berkeley, R.C.W., Madelin, M.F. and Round, F.E. (1971) *Microorganisms Function, Form and Environment* (L.E. Hawker and A.H. Linton eds). Edward Arnold, London, p. 275.

74. Burnett, J.H. (1968) *Fundamentals of Mycology.* Edward Arnold, London.

75. Alberts, B., Bray, D., Lewis, J., Raff, M., Roberts, K. and Watson, J. (1989) *Molecular Biology of the Cell (2nd edn).* Garland Publishing, New York.

76. De Greef, J.A. (1991) *Microscopy and Analysis*, **26**, 25.

7. APPENDIX: SUPPLIERS

7.1. General dyes and stains

Aldrich Chemical Co. Ltd., The Old Brickyard, New Rd., Gillingham, Dorset SP8 4JL, UK.
BDH Chemicals Ltd., Poole, UK (also suppliers of Gurr products).
Fison's Scientific Equipment, Bishop Meadow Rd., Loughborough LE11 0RG, UK.
National Diagnostics, 303 Cleveland Ave., Highland Park, New Jersey 08904, USA.
Sigma Chemical Co. Ltd., Fancy Rd., Poole, Dorset BH17 7NH, UK.

7.2. Specialist fluorescent probes

Molecular Probes Inc., 4849 Pitchford Ave., Eugene, USA (UK supplier: Cambridge Bioscience, 25 Signet Court, Stourbridge Common Business Centre, Cambridge CB5 8LA).

7.3. Immunochemicals

Amersham International PLC, Northern Europe Region, Lincoln Place, Green End, Aylesbury HP20 2TP, UK.
Dako Ltd., 16 Manor Courtyard, Hughenden Ave., High Wycombe, Bucks HP13 5RE, UK.
Pierce Chemical Co., 3747 North Meridian Rd., PO Box 117, Rockford, IL 61105, USA.
Serotec, 22 Bankside, Station Approach, Kidlington, Oxford OX5 1JE, UK.
Sigma Chemical Co. Ltd., Fancy Rd., Poole, Dorset BH17 7NH, UK.
Vector Labs. Inc., 30 Ingold Rd., Burlingame, CA 94010, USA.

CHAPTER 4
ELECTRON MICROSCOPY

N. Harris

An increasing array of approaches and techniques is available to the cell biologist wishing to use electron microscopy not just to determine fine structural detail but also to obtain specific molecular, biochemical and physiological information, which can be combined to give a greater understanding of spatial and temporal aspects of processes occurring within plant and animal cells.

As with all of the techniques applied to cell biology, those relating to the preparation and examination of specimens by electron microscopy need to be 'tuned' for each particular type of specimen and investigation. All 'good' electron microscopists spend quite a lot of time looking at their (preferably living) cells and tissues under an optical microscope!

Many of the reagents used in electron microscopy are toxic or very toxic, carcinogenic or very carcinogenic, and/or teratogenic. Good laboratory practice and appropriate safety precautions are essential.

1. SPECIMEN SUPPORTS FOR TRANSMISSION ELECTRON MICROSCOPY (TEM)

A wide variety of specimen grids are available. Generally a metallic mesh is used; the material, size and type of mesh, and coating should be selected with regard to the specimen and type of study being undertaken. The major variables are listed in *Table 1.*

Formvar films are made from 0.3–0.8% Formvar in chloroform (1). A clean slide is dipped into the solution and dried carefully to avoid condensation of very small water droplets; these produce holey film. Film is floated on to clean water, grids are laid on to the film and then picked up using either a glass slide or, more easily, a piece of partially absorbent paper.

Holey or lacey films are made by encouraging condensation as the film dries, or by including 1–10% water in the dipping solution; these films must be stabilized with carbon.

Glow discharging is used to reduce the hydrophobicity of carbon films/coating and facilitate uniform spread of aqueous suspensions and stains.

2. NEGATIVE STAINING OF MACROMOLECULES AND SMALL OBJECTS

Developed during the late 1950s, this technique has proved very valuable for examining fine surface details of small, particulate specimens, such as viruses and some macromolecules (1). It has also been used more recently to visualize spreads of double-stranded (ds) nucleic acids. Single-stranded nucleic acids require rotary shadowing for visualization (2).

Negative staining is quick and reliable, it is appropriate for routine screening of a wide range of samples. Specimens are deposited on to coated grids and briefly stained with an electron-dense, aqueous stain; excess stain is washed away leaving marked shape and detailed surface features. Commonly used negative stains are detailed in *Table 2.*

Table 1. Specimen grids for TEM

Material	
Copper	This is used with standard staining; one-sided coating (e.g. rhodium) helps in extended series of stainings (e.g. for immunocytochemistry)
Gold/nickel	These are required when etching resin sections or using corrosive histochemical reagents; gilded/platinized copper grids are more robust than pure gold
Light-element (e.g. beryllium, aluminum or carbon composite)	These reduce background in elemental detection and microanalysis by X-rays
Pattern	
Mesh	These have square, rectangular or hexagonal holes; mesh is rated in holes per inch, standard 75 has holes of approx. 320 μm; 200 size (100 μm holes) is popular; hexagonal style has maximum open area; use rim mark for orientation
Narrow-bar	These give higher transmission (e.g. 85% for a 200 mesh cf. 60% for standard 200); standard are more robust
Slot and parallel bar	These are used for serial sections; practice is required to pick up ribbon along slot
Finder	These have a variety of patterns to identify specific grid squares; useful where specimens are re-examined
Folding	Folding grids may be used for 'thick' (> 0.2 μm) sections
Coatings	
Uncoated	Uncoated grids may require a brief rinse in 1M HCl and distilled water, to etch the surface and facilitate section adhesion
Carbon	Carbon-only film is used for magnifications in excess of 100 000–150 000
Plastics	Plastic coatings such as Formvar or Parlodion are used routinely to enable viewing of greater section areas, particularly with wider mesh grids, but they reduce obtainable resolution and are not very stable under high beam irradiance; stability is improved by carbon coating
Holey/lacey	These give partial support, for example to sections or crystals, while allowing observation of regions of the specimen without interference from the film

3. CONVENTIONAL THIN-SECTION STUDIES FOR TEM

Ultrastructural detail of cell contents is usually demonstrated by examination of ultrathin (0.08–0.1 μm thick) sections. Biological samples cannot be cut this thinly without a matrix for structural support. The matrix is usually resin, but more recently techniques utilizing the rapidly frozen cell sap have been introduced. Generally, appropriate resins are not able to

Table 2. Negative stains

Phosphotungstic acid/sodium phosphotungstate This stain is used as a 0.5–2% aqueous solution (usually 1%); add drops of conc. NaOH for desired pH (usually 5–6 for viruses and macromolecules); a 'wetting' agent may be useful, e.g. > 0.1% serum albumin. The staining time used is 0.5–5 min (usually 1 min)
Uranyl acetate This is used as a 0.5–2% aqueous solution (usually 1%); pH 4.3 or further acidified; the staining time used is 0.5–5 min (usually 1 min)
Uranyl formate and uranyl nitrate These are used in the same way as uranyl acetate, but are generally less popular
Calibration standards Viruses, e.g. TMV at 0.03 mg ml^{-1}, may be included

permeate freely into fresh (aqueous) biological tissue. Tissue preparation is thus based on three-step protocol:

(i) fixation to retain structure in a relatively 'lifelike', although now static form;
(ii) dehydration;
(iii) infiltration and polymerization of resin.

Samples should preferably be no more than 1–2 mm in at least one dimension. Use solution volumes of at least 10× (and better 50×) that of the sample. Gentle agitation is advantageous; rotators which take capped, 10 ml specimen vials are very convenient, and readily available commercially.

3.1. Fixation of samples

Ideally the primary fixative should match the pH, osmolarity and ionic concentration of the cell sap; buffering capacity is also important (3). Buffer can have a marked effect on the final image. Phosphate (0.05–0.1 M, pH 6.0–7.5), cacodylate, and Pipes buffers are common. Phosphate is generally considered the 'most physiological'. Cacodylate contains arsenic, which ensures a long shelf life but it should only be used in a fume hood (the pH is adjusted with HCl). Pipes buffer appears to give a greater retention of cytosolic and endoplasmic reticulum (ER) enclosed proteins. The temperature of the primary fixation (usually 4°C or 20°C) may influence fixation and retention of specific components (e.g. microtubules). Common fixatives for TEM are listed in *Table 3*.

Buffered aldehydes are the standard primary fixatives for TEM thin sections (3), and thick sections (4). Concentrations and relative proportions of glutaraldehyde and (*para*)formaldehyde depend on the primary aim of the investigation: for cytology, higher concentrations (e.g. 4% exclusively, or 2.5%) of glutaraldehyde are used with lower concentrations (e.g. 1.5%) of (*para*)formaldehyde; for immunocytochemical work (5), primary fixation is with 2.5–4% (*para*)formaldehyde with or without 1.5% glutaraldehyde.

The extent of fixation is dependent upon the type of specimen and its size; typically 1–4 h for small pieces of soft animal tissue, up to 24 h for larger pieces of plant tissue. Be sure to optimize fixative and buffer combinations and concentrations before starting on an extensive study; such optimization may be essential for immunocytochemistry.

Table 3. Fixatives for TEM

Aldehydes

These cross-link proteins and nucleic acids mainly by reacting with amino groups

Glutaraldehyde

This is used at 0.5–4%, buffered solution, for 1–16 h at 20°C or 4°C; often used in combination with (*para*)formaldehyde

(*para*)Formaldehyde

This is used at 0.5–4%, buffered solution; stock made freshly from crystalline form by dissolving 20% w/v in water at 65°C and adding drops of 1 M NaOH until the suspension clears

Acrolein (acrylic aldehyde)

This is extremely toxic; it penetrates tissue very rapidly but is seldom used now other than for large pieces of (plant) tissue

Osmium tetroxide

This is a strong oxidizing agent which fixes and also stains unsaturated lipids; its solution and vapor are extremely toxic and should be used with great care only in fume hood; it is used at a concentration of 0.5–2%, usually aqueous but may be buffered; treatment is for 1–16 h and generally as a secondary fixative after an aldehyde primary; if used as the primary, there is a high loss of proteins and RNA

Potassium permanganate

This is used as a 1% aqueous solution; it gives high-contrast images of membranes; however, considerable care should be taken when interpreting images, which are prone to show a range of artifacts, including dilated organelles and nuclear pores, collapsed (plant) vacuoles, etc.; it is used less frequently nowadays

Zinc–iodine–osmium tetroxide (ZIO)

This is used for enhanced selective staining of the lumen of endoplasmic reticulum, Golgi, nuclear envelope, prolamelar bodies, etc.; better results are ususally obtained with primary fixation (4)

Use freshly made; it is made by adding 1.5 g zinc powder and 0.5 g iodine to 10 ml water; it is mixed thoroughly and filtered after 5 min, and the solution is combined with an equal volume of 2% aqueous osmium tetroxide; it is used for 6–18 h at 20°C

Freezing

This is a temporary immobilization rather than true fixation, but it is becoming increasing popular as it helps to avoid chemical treatment of the tissues (see Section 9.1)

3.2. Dehydration

Dehydration of samples is carried out via a graded series of alcohol or acetone treatments. The starting point and the number and duration of steps depend upon the size and type of sample, although typically an alcohol dehydration series would use 10%, 25%, 50%, 75%, 95% and 100% alcohol (in water or buffer) with one to three incubations at each stage for 15–30 min. Propylene oxide is used as an intermediary between alcohol and Araldite™ and Epon™ resins.

3.3. Infiltration

The infiltration of samples with resin is done by incubations with stepwise increases in resin concentration in the dehydrating reagent, typically 30%, 60% and 100% resin with several changes of pure resin before polymerization.

3.4. Embedding samples in resin

Essential characteristics for resins include a low viscosity, to aid infiltration; uniform polymerization, which should involve little or no change in volume; uniform sectioning characteristics and stability in the vacuum of the electron microscope. Mixtures of the resin components, usually the resin, a hardener, a plasticizer and the accelerator for polymerization, can be altered to vary the cutting characteristics. Commercial suppliers of EM consumables will supply individual components or kits with pre-aliquoted components to be mixed for 'hard', 'medium' or 'soft' blocks. Commonly used resins and their component mixtures are listed in *Table 4.*

3.5. Sectioning of embedded samples

After resin polymerization, sections are cut using ultramicrotomes with glass or diamond knives. Freshly cleaved glass knives are relatively cheap, have an effective cutting edge of a few millimeters but lose their 'edge' in each region after cutting 20–30 sections. This necessitates repositioning of the knife edge. Diamond knives are initially rather more expensive but, if maintained properly, have a much more durable edge and are ideal for serial sectioning or cutting numerous very thin sections.

Section thickness may be gauged by the interference color of the section (see *Table 5*) as it lies on the water trough after cutting.

Staining improves contrast in sections for examination in the TEM (6). Salts of heavy metals are used, most commonly uranyl acetate (at 5% to saturated aqueous solutions) and/or Reynold's lead citrate. (Dissolve 1.33 g lead nitrate in *c.* 30 ml distilled water in a 50 ml volumetric flask and add 1.76 g sodium citrate; shake the resultant suspension vigorously every few minutes for 30 min and add 8 ml freshly made 1 M NaOH. The suspension should clear, leaving a solution of alkaline (*c.* pH 12) lead citrate. Make up to 50 ml and use or store for up to 6 months.) Typical staining times are 15–30 min for uranyl acetate, and 20–60 min for lead citrate.

3.6. Examination of sections by TEM *(Table 6)*

The image contrast varies with accelerating voltage (higher contrast at 40–60 kV compared with 80–100 kV).

Image brightness varies with the filament currents and condenser focus (higher current reduces filament life; micrographs should be taken with a defocused condenser to avoid a central 'hot spot')

If the microscope has a tilt stage, use it, because the 'sharpness' of images of, for example, stacks of membranes and crystalline lattices may be improved substantially.

Take micrographs at one magnification stop below the 'framed' image you might consider publishing (this allows for trimming and plate montages).

4. THICK-SECTION STUDIES (high voltage microscopy is not always needed)

At approximately 80 nm, 'thin' sections are essentially two-dimensional with regard to many cellular organelles and structures. The third dimension may be determined either by examination of serial sections and reconstruction of appropriate structures or, more directly, by examining thicker sections.

Table 4. Embedding media

Epoxy resins

Aromatic, e.g. Araldite™

This medium is used extensively for animal and plant tissues, it has high stability; variation in components alters the characteristics for sectioning; a standard mix contains 19.0 ml Araldite™ CY212, 21.0 ml DDSA hardener, 0.6 ml dibutyl phthalate plasticiser, 1.2 ml BDMA accelerator; it is polymerized at 60°C for 24 h

Aliphatic, e.g. Epon 812 substitutes

This type of medium is also used extensively; it has lower viscosity than aromatics but is less stable in the electron microscope; it requires a second hardener; there are a number of variations of the original formulation; the standard mix contains 20 ml Epon™, 16 ml DDSA, 8 ml MNA, 1.3 ml BDMA; it is polymerized at 60°C

Vinyl cyclohexane dioxide based, e.g. Spurr

These are used extensively, particularly for plant tissues; handle them with care as one component is carcinogenic, so it is best purchased as a kit with pre-aliquoted components

Acrylic resins

Lowicryl K4M and HM20

These media are useful for low-temperature embedding; K4M allows infiltration and u.v. (360 nm) curing at −35°C, and HM20 at −70°C; thermal polymerization can also be used; K4M is a water-compatible polar resin, HM20 is hydrophobic and non-polar; these resins are very popular for immunocytochemical and *in situ* hybridization studies

LR White

This is a low-viscosity, hydrophilic resin, which can be cured chemically or at 65°C in completely filled, air-impermeable (e.g. polypropylene) capsules, stored at 4°C; uneven polymerization may occur after osmium postfixation; the resin is very popular for immunocytochemical, and combined light and electron microscope studies; its low viscosity permits infiltration of woody plant and decalcified bone specimens

LR Gold

This medium is designed primarily for embedding unfixed tissues at low temperatures (PVP is used to protect unfixed tissues from osmotic damage); it is used for sensitive enzymes, immunocytochemistry and *in situ* hybridization; the resin can be cured with initiator and visible light

Other methacrylates

These are referred to widely in older literature but they are seldom used now as they result in considerable specimen shrinkage and are unstable in the electron microscope; water-soluble versions have some continued use for specialized cytochemical studies

Abbreviations: PVP, polyvinylpyrrolidone; DDSA, dodecenyl succinic anhydride; BDMA, benzyl dimethylamine; MNA, methyl nadic anhydride; CY212, diglycidyl ether of bis-phenol A.

Sections up to 10 μm thick can be examined in high-voltage (200–3000 kV) microscopes, but biological tissues can give substantial three-dimensional information from conventional (80–120 kV) microscopes, particularly if the samples have been selectively stained.

Table 5. Section thickness for TEM

Appearance of sections	Thickness	Comments
Translucent gray/silver	60–90 nm	Gives very 'crisp' images
Pale yellow	90–120 nm	Usable and easier for beginners to cut
Gold	120–150 nm	May be used for work at lower magnifications ($<$10 000)
Iridescent purple/green	150–250 nm	For 'thick' section studies at 100 kV
Opaque silver	$>$0.5 μm	For high-voltage electron microscopy (HVEM) (and light microscopy with sections stained using 0.1% toluidene blue in 0.1 M sodium tetraborate)

Table 6. Typical shapes and dimensions of some ultrastructural features

Viruses	Isodiametric 50–100 nm diam.; rod 30 nm × several μm
Bacteria	Spherical, rod or spiral; 1–10 μm
Fungi	Unicellular ($>$5 μm) or hyphal (2–10 μm diam.) eucaryotic; hyphae with simple (0.05–0.5 μm) or doli-pores in cross-septa
Organelles of eucaryotic cells	Nucleus: typically spherical; 3–30 μm diam.; double-membrane envelope with pores (pore complex 70–90 nm diam.) Chloroplast: plano-convex disc; 5–10 μm diam.; double-membrane envelope; inner membranes (grana and thylakoids) not attached to envelope; starch and lipid inclusions; various other plastid types Mitochondria: simple rod to complex multifingered; 0.5 μm diam., variable length; double-membrane envelope, cristae extensions of inner membrane Golgi apparatus: complex of flattened cisternae and associated vesicles; overall diam. 0.7–2 μm; stack of cisternae appears highly polarized in plant cells, generally less so in animal cells Vesicles and vacuoles; 60 nm diam.; to very large (plant) vacuoles; single bounding membrane; small vesicles may be coated with clathrin or non-clathrin protein; wide range of inclusions, depending upon cell type and role of vesicle or vacuole Microbodies (peroxisomes/glyoxysomes): spherical; 0.2–2 μm diam.; single bounding membrane, may have crystalline inclusion(s) Ribosomes: 20 × 30 nm; cytosolic or membrane bound

The three-dimensional interrelationships of endomembrane system components are shown in thick (0.3–1 μm) sections of tissue fixed with either ZIO (4) or potassium ferricyanide, examined at 100 kV.

Use a tilting, and ideally rotating and tilting, specimen stage for thick sections to distinguish between real interconnections and apparent ones resulting from overlay within the thickness of the section. Micrographs to distinguish such features may be produced as either high-tilt or stereo pairs.

High-tilt pairs use angles of 30° and 60° between micrographs; overlapping points are displaced laterally but interconnected points are not.

The tilt angle for stereo pairs is determined by:

$$\Delta H = P/2M_t \sin\omega$$

where ΔH is specimen thickness; P, parallax; M_t, total magnification; and ω is tilt angle (7). Stereo pairs can be viewed as micrographs or projected images.

5. EM AUTORADIOGRAPHY *(Table 7)*

EM autoradiography was the first technique to combine the power of biochemical analysis with fine spatial resolution. It has proved very valuable in, for example, pulse-chase studies of precursor incorporation, and has been the subject of a number of books and reviews (e.g. ref. 8). In principle, the technique is reasonably straightforward, although technically a little demanding.

After a suitable exposure time (determined empirically) the emulsion is processed to develop and fix the affected silver grains. The signal can be produced as either 'worm-like' tracks (e.g. using Ilford L4 emulsion with Microdol X developer) or spherical grains (e.g. Ilford L4 with gold latensification and Agfa-Gevaert fine-grain developer); 20% sodium thiosulfate is used as fixative.

If contrast staining of the specimen is required, this may be done with uranyl acetate and lead citrate after first removing the gelatin film (by floating the grids on a water bath at 37°C for 20 min).

Table 7. Isotopes and emulsions for EM autoradiography

Isotopes
The choice is determined by the required resolution and available time; ^3H gives best resolution but requires longer exposure times, ^{32}P gives poor resolution but much shorter exposure times, ^{35}S is intermediate in both but more expensive and often it is not available in as wide a range of precursors
Emulsions
A changing range, specifically formulated for EM autoradiography, is available commercially; select for the particular isotope and use carefully
Non-specific signals
These arise from: poorly stored emulsion; use of dirty or metallic forceps; excess 'safe-light' levels; or 'dirty' solutions and vessels used for emulsion development (very clean vessels, solutions made with deionized water and carefully matched for temperature, are important)

6. IMMUNOCYTOCHEMISTRY FOR EM *(Tables 8 and 9)*

Immunocytochemistry is used to locate specific proteins within samples; it is very effective in ultrastructural studies (9).

'Etching' of sections is used to increase the level/density of immunolabeling when samples have been embedded in epoxy resins; sections must be collected on inert (e.g. gold) grid.

Table 8. Outline for single immunolabeling

(i) Sections are mounted on to coated, copper grids (200 mesh) if no resin 'etching' is required, or gold grids if etching is to be used (epoxy sections may be etched with aqueous saturated sodium metaperiodate for 30 min)

(ii) Block non-specific antibody binding (using e.g. 0.1–1% bovine serum albumin)

(iii) Incubate with primary antibody for 30–60 min; controls include use of pre-immune serum at similar dilution

(iv) Wash thoroughly

(v) Incubate in secondary antibody conjugated to a marker

(vi) Wash thoroughly, contrast stain and examine in EM

Table 9. Markers for EM immunolabeling

Colloidal gold conjugated to secondary antibody
- This is very popular and a wide range of conjugated secondary antibodies are available commercially
- Colloid sizes range from 1 to 25 nm
- The labeling density is inversely proportional to colloid size
- Sizes less than 10 mm are difficult to see in lower-power electron micrographs

Colloidal gold conjugated to Protein A
- The physical characteristics and advantages are as above, but Protein A does not bind to all primaries

Conjugated ferritin
- This was used in early EM immunostudies but is now supplanted by colloidal gold

Enzyme conjugates
- These are popular for light microscopy (LM) but rarely used for EM studies as they give comparatively poor resolution

Double immunogold labeling
- A minimum of 10 nm difference between the two colloids is recommended
- As labeling density is dependent upon colloid size, experiments must include each combination of primary with secondary conjugate

A variety of proteins are used to block non-specific binding of the primary and secondary antisera, including bovine serum albumin and pre-immune serum; the optimal times and concentrations are determined empirically.

The concentration and incubation times for the primary antibody are optimized using serial dilutions to determine conditions for maximum density of specific labeling without increasing background, non-specific labeling. In practice, a wide variety of dilutions, incubation times and temperature (e.g. 4°–30°C) are used for different combinations of primary antisera and embedded-tissue antigen. A detailed consideration of these factors is given in ref. 9.

The size of the gold colloid to be used should be selected carefully; small colloids (e.g. 5 nm) give higher levels of labeling but are 'less visible' at the lower magnifications that may be necessary for giving an overview of immunolabeled product. Larger colloids (e.g. 20 nm) are more readily visible but give lower levels of labeling. A compromise is the use of a heterodispersive colloid which includes large and small particles: the image is, however, perhaps not as aesthetically attractive for publications. The use of heterodispersive colloids also saves the extensive centrifugations required for preparing uniformally sized monodispersive colloids. Such, mixed-size colloids can be made 'at home', and can be combined to the secondary antibody or Protein A quite simply and relatively cheaply (9).

7. *IN SITU* HYBRIDIZATION FOR EM

Localization of specific DNA and RNA sequences by *in situ* hybridization can be undertaken at the EM level. Various approaches are possible and the main choices are outlined in *Figure 1*.

The choice of when to section and when to hybridize may be determined largely by the nature of the tissue being studied (10). For cell monolayers, epidermal cells, suspension cells and plant protoplasts a 'mild' fixation, hybridization, staining, embedding and sectioning may be suitable. For cells within a multicellular mass, however, 'mild' fixation, sectioning of the tissue (to, for example, 10–100 μm), hybridization, staining, embedding and thin sectioning may be more appropriate. It is possible to monitor the hybridization reaction by light microscopy (LM) and then select appropriate samples for subsequent EM examination, although this does require that the LM sample has been prepared with this subsequent step in mind.

Figure 1. Protocol choices for EM *in situ* hybridization.

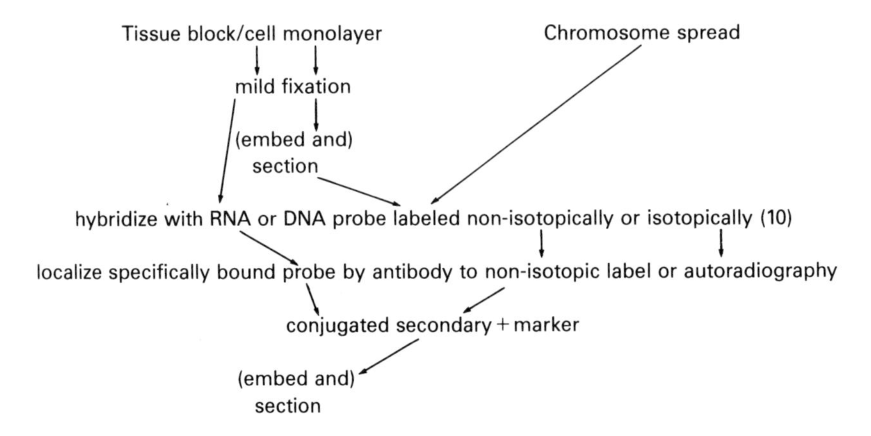

Table 10. Probes, labeling and localization for EM *in situ* hybridization

Probes		
DNA	usually as ds (except M13); or oligonucleotide	Typical probe lengths 200–600 bp, except oligos where 'cocktails' of shorter probes may be useful
RNA	single strand (sense strand is excellent control)	DNA probes are more 'robust' than RNA but RNA–RNA hybrids are more stable
Labeling		
Isotopic	Generally ^3H	Localized by autoradiography
Non-isotopic	Biotin	Localized by antibiotin and secondary + marker[a]; avidin/streptavidin + marker[a]
	Digoxygenin	Localized by antidigoxygenin and secondary + marker[a]
	Others (11)	

[a] Markers include: colloidal gold which is popular and very effective; ferritin (supplanted by colloidal gold); enzyme + cytochemical localization (for in-block localizations).

Electron Microscopy

The choice of probe labeling is affected by the required resolution of distributional detail; the path length of isotope tracks in the monolayer emulsion of EM work may be too long for some studies, while the 'visibility' of smaller probe-labels, such as ferritin, may be unsuitable for distributional analyses of label at relatively low magnifications. A further consideration is the penetration of the marker into the tissue. Particulate markers may require greater disruption of the tissues than soluble stain reactants which must produce an insoluble product.

In situ hybridization reactions can be carried out directly on samples (chromosome spreads (12) or sections (13)) on EM grids.

Double-labeling *in situ* hybridization at EM level is carried out using, for example, one probe labeled with biotin and one with digoxygenin; the bound probes are localized with different-sized gold colloids (13).

As with other cytochemical techniques, for best results it is essential that protocols are 'tuned' to each specific tissue and probe combination.

8. SCANNING ELECTRON MICROSCOPY (SEM)

Scanning electron microscopy is generally used for looking at the surfaces of specimens, and consequently the careful retention of internal cellular detail may not be so critical, unless it has a marked influence on the surface features of interest, or the surface being examined is an internal face exposed by fracturing.

For samples with hard surface features, which are not distorted by the vacuum environment within the electron microscope, there may be very little specimen preparation beyond selecting an appropriately sized sample and, if it is non-conducting, coating with a thin conductant film which allows for removal of any excess charge. However, biological samples often have water as a part of their structural integrity, in which case the water must be 'removed' without

Table 11. Sample preparation for SEM

Air-drying Perhaps the easiest sample preparation for EM; used where the specimen has a surface structure which is not affected by loss of water, such as some exoskeletons, pollen grains, etc.; the sample is simply left to dry in a clean environment in the presence of a desiccant, and, optionally, with vacuum assistance to speed up the process; when dry it is ready to be mounted onto a specimen stub and, if required, coated before examination in the SEM
Dry fracturing (Vapor) fixation before fracturing may influence the fracture plane
Freeze-drying Used as an alternative to air-drying
Critical-point drying (14) Used to remove water from samples that are essentially 'soft', but without the sample collapsing; this is suitable for animal cells and tissues, soft plant tissues, protoplasts, etc.
Cryo-SEM (15) This is particularly valuable for biologists; chemical fixation may not be needed, frozen internal fluids provide the same structural support and shape as *in vivo*; *it is the only method for examining surface liquids and exudates* in situ. Some cells survive EM examination (16)

distorting the specimen's surface. This may be achieved by critical-point drying, or holding the water in a frozen state even when the specimen is within the microscope (*Table 11*).

Samples are examined in the SEM after mounting onto a stub. Aluminum stubs are often used although carbon may be required for analytical work where background radiation is a problem.

The sample is attached to the stub either with electron-conductive adhesive (e.g. silver paint or colloidal graphite) or double-sided sticky tape.

Biological specimens are generally not electron-conducting, without a coating they become charged and very inferior images are obtained. Specimens are usually coated with a thin (9–15 nm) metallic (gold or gold/palladium) film produced in a Sputter Coater. The process is quick and simple, and most commercially available coaters can cope with six or more stubs at a time.

9. OTHER APPROACHES: A BRIEF AND SELECTIVE OVERVIEW

9.1. Cryo-EM

Water is an integral part of cell biology; a major feature of cryo-SEM is the ultrastructural examination of hydrated samples from, for example, viruses (17) to secreting glands (18). The rate of freezing is critical to the production of vitreous water (ice damages specimens); cooling rates in excess of 10^2 K per second are required for biological specimens.

Freeze fixation
This is really only immobilization and is used to avoid chemical fixation; it may be followed by:

Freeze drying (15).

Fracturing and etching. Sublimation of some water to enhance surface features; for SEM or surface-replica production for TEM examination (15); deep etch (19) (for cytoskeleton, clathrin vesicles, etc.).

Cryo-ultramicrotomy. To produce sections for immunocytochemistry (20) or analysis of elemental distributions (21).

Freeze substitution. And the production of sections of resin-embedded tissue (15).

The field of cryo-EM is generally dominated by high-technology, specialist laboratories; however, cryo-SEM is a commercially available add-on, and freeze substitution does not require specialist equipment if the sample is amenable to freezing by plunging into liquid nitrogen (e.g. unicellular eucaryotes).

9.2. Elemental analysis
This may be done by the analysis of X-rays emitted from the electron-irradiated sample, using characteristic X-ray energies of wavelengths from each element (21). Energy-detection systems are available in most material sciences research groups.

An alternative method is by analysis of electron energy loss. This is less commonly available but may be more appropriate for biological studies as it is better suited to lighter ('physiologically interesting') elements.

9.3. Scanning transmission electron microscopy
This combines TEM with SEM. It gives high-resolution SEM, substantial advantages for examining thick sections, and assists with quantitative mass and element mapping at high resolution (22).

9.4. Low-dose techniques
These use a special feature of some electron microscopes which allows examination of samples that are normally very sensitive to damage by electron-beam bombardment. They can be useful in the image analysis of molecular structure (23).

10. REFERENCES

1. Pease, D.C. (1964) *Histological Techniques for Electron Micrsocopy.* Academic Press, New York.

2. Sommerville, J. and Scheer, U. (1987) *Electron Microscopy in Molecular Biology: A Practical Approach.* IRL Press, Oxford.

3. Glauert, A. (1975) *Fixation, Dehydration and Embedding of Biological Specimens.* North-Holland, Amsterdam.

4. Harris, N. (1978) *Planta*, **141**, 121.

5. Beesley, J.E. (1992) *Immunocytochemistry: A Practical Approach.* IRL Press, Oxford.

6. Lewis, P. R. and Knight, D. P. (1977) *Staining Methods for Sectioned Material.* North-Holland, Amsterdam.

7. Nankivell, J.F. (1963) *Optik*, **20**, 171.

8. Williams, M. A. (1977) *Autoradiography and Immunocytochemistry.* North-Holland, Amsterdam.

9. Polak, J.M. and Varndell, I.M. (1984) *Immunolabelling for Electron Microscopy.* Elsevier, Amsterdam.

10. Harris, N. Wilkinson, D.G. (1990) In situ *Hybridisation: Application to Developmental Biology and Medicine.* Cambridge University Press, Cambridge.

11. Coulton, G. (1990) in In situ *Hybridisations: Application to Developmental Biology and Medicine* (N. Harris and D.G. Wilkinson eds). Cambridge University Press, Cambridge, p. 1.

12. Hutchinson, N. (1984) in *Immunolabelling for Electron Microscopy* (J.M. Polak and I.M. Varndell, eds). Elsevier, Amsterdam.

13. McFadden, G.I. (1991) in *Electron Microscopy of Plant Cells* (J.L. Hall and C. Harves, eds). Academic Press, London.

14. Borrelli, M.J., Koehm, S., Cain, C.A and Tompkins, W.A.F. (1985) *J. Ultrastr. Res.*, **91**, 57.

15. Robards, A.W. and Sleytr, U.B. (1985) *Low Temperature Methods in Biological Electron Microscopy.* Elsevier, Amsterdam.

16. Read, N.D. and Lord, K.M. (1991) *Exp. Mycol.*, **15**, 132

17. Adrian, M., Dubuchet, J., Lepault, J. and McDowall, A.W. (1984) *Nature*, **308**, 32

18. Robards, A.W. (1985) in *Botanical Microscopy* (A.W. Robards ed). Oxford University Press, Oxford, p. 39

19. Heuser, J.E. (1980) *J. Cell Biol.*, **84**, 560

20. Tokyasu, K. (1986) *J. Microscop.*, **143**, 139

21. Morgan, A.J. (1985) *X ray Microanalysis in Electron Microscopy for Biologists.* Oxford University Press, Oxford.

22. Crewe, A.V. (1984) *J. Ultrastr. Res.*, **88**, 94

23. Ohtsuki, M. and Crewe, A.V. (1983) *J. Ultrastr. Res.*, **83**, 312.

11. APPENDIX

Some of many suppliers of equipment and consumables for EM sample preparation with world-wide distribution services

Agar Scientific Ltd., 66a Cambridge Road, Stansted, Essex CM24 8DA, UK.
Baltzers, Pf75, FL-9496 Balzers, Liechtenstein.
Bio-Rad, Watford Business Park, Watford, Herts WD1 8QS, UK.
Bio-Rad, 19 Blackstone Street, Cambridge, MA 02139, USA.
Polysciences, 400 Valley Rd, Warrington, PA 18976, USA.
Taab Laboratories Equipment Ltd., Calleva Industrial Park, Aldermaston RG7 4QW, UK.

Isotopic and non-isotopic labeling and detection systems
Amersham, Aylesbury, Bucks HP20 2TP, UK.

Biotin. Vector, 30 Ingold Rd, Burlingame, CA 94010, USA.
Vector, 16 Wolfric Square, Peterborough PE3 8RF, UK.
Digoxygenin. Boehringer Mannheim, D-6800 Mannheim 31, Germany.

Cryo-EM and cryo-SEM
Oxford Hexland, Eynsham, Oxford OX8 1TL, UK.

CHAPTER 5

DNA AND CHROMOSOMES OF CELLS

D. Patel and D. Rickwood

Table 1. DNA and chromosomes of cells: bacteria

Organism	DNA/cell (pg)	Mole % G + C[a]	Unique % DNA	Refs
Acetobacter				
ascendens		55(T)		1
mobilis		58.9(T)		2
Actinomyces				
badius		70(T)		3
fluoroscens		73(T)		3
israelii		60.3(T)		4
Agrobacterium				
rhizogenes		62.8(T)		5
tumefaciens		58(T)		6
Alcaligenes				
denitrificans		69.8(T)		7
faecalis		54.8(T)		8
Arthrobacter				
aurescens		62.9(T)		9
globiformis		60(T)		10
Azotobacter				
agilis		53.5(T)		11
beijerinckii		66.2(T)		11
chroococcum		64.8(T)		11
Bacillus				
cereus (spores)	0.0108	36(D)		12[b], 13
(veg. cells)	0.0129	36(D)		14, 13
licheniformis		46(T)		15
megaterium	0.025	45.5(C)		16, 17
subtilis		44.5(T)		18
(spores)	0.00542			12[b]
ws (veg. cells)	0.0048			19
Clostridium				
acidiurici		25.3(T)		20
pasteurianum		30.5(T)		20
perfringens		26.5(T)		21
Desulfovibrio				
aestuarii		55(D)		21
orientis		41.7(D)		22

Table 1. Continued

Organism	DNA/cell (pg)	Mole % G + C[a]	Unique % DNA	Refs
Escherichia				
coli	0.009	51.5(T)		23, 24
B (log phase)	0.0137			25
(stat. phase)	0.0078			25
Flavobacterium				
capsulatum		63(T)		26
odoratum		35(T)		27
suaveolens		67(T)		8
Klebsiella				
aerogenes		53(C)		27
pneumoniae		59.2(T)		28
Lactobacillus				
acidophilus		34.9(C)		29
casei		47(T)		30
Leuconostoc				
mesenteroides		39(T)		21
Nitrobacter agilis		65(D)		31
Nitrosomonas europaea		50(T)		32
Pseudomonas				
alcaligenes		66.3(D)		33
denitrificans		57.5(T)		34
fluorescens		60(T)		21
Rhizobium				
japonicum		61(T)		35
leguminosarum		61.4(T)		36
meliloti		62.6(C)		29
Salmonella				
typhimurium	0.011			37
typhi (typhosa)		50(T)		38
Shigella dysenteriae		53.4(C)		39
Streptococcus				
pyogenes		36.5(C)		39, 40
thermophilus		30.5(T)		41
Streptomyces				
albus		71.4(D)		42
caeruleum		72(T)		43
griseus		71(C)		3
Thiobacillus				
denitrificans		64(D)		32
ferroxidans		57(D)		32
Vitreoscilla species		44(D)		44
Xanthomonas phaseoli		66.4(D)		45

[a] Mole % guanosine and cytosine (G + C) was determined by one of the following methods, which are given in their abbreviated form in the table: T_m, T; chemical analysis, C; buoyant density, D. Values are given to the nearest 0.1%.

[b] Content of DNA estimated from the phosphorus analysis cited in the reference and a value of 9.23% phosphorus in DNA.

Table 2. DNA and chromosomes of cells: protozoa, algae, fungi, echinoderms and arthropods

Organism	Ploidy[c]	DNA/cell (pg)	Mole % G + C[a]	Unique % DNA	Refs
Protozoa					
Euglena gracilis	?x = 45	2.9			46, 47
var. *bacillaris*			50.2(T)[d]		48
Paramecium					
aurelia[e]			29(D)	85	49, 50
caudatum		1.53–1.67	28.2(C)		51, 52
Tetrahymena pyriformis		13.6	25(C)	80	50, 53, 54
Trichomonas vaginalis			29(D)		55
Trypanosoma equiperdum	?x = 3		47.9(D)		56, 57
Algae					
Anabaena variabilis		0.036			58
Anacystis nidulans		0.030			58
Chylamydomonas reinhardi			64(D)[d]	70[f]	59, 60
Fungi					
Aspergillus nidulans	x = 8		47(C)		61, 62
Green haploid		0.0438			
(per conidium)					63[b]
Candida albicans			32.5(T)		64
Dictyostelium discoideum			22(C)		21
whole cell DNA				40[f]	65
nuclear DNA				60[f]	65
Neurospora crassa	x = 7		55(T)	80	66, 67, 68
Physarum polycephalum (slime mould)	x = 8		42.5(D)[d]	58[f]	69, 70, 71
Saccharomyces cerevisiae	x = 16	0.046	39.5(T)		72, 73, 74
haploid		0.0245			75[b]
Echinodermata					
Sea urchin					
Lytechinus, sperm		0.90			76
Paracentrotus, sperm		0.70	36.1(C)		77, 78
Strongylocentrotus purpuratus				38	79, 80
Arthropoda					
Aedes caspius, spermatid		0.988			81
Artemia salina	2x = 42				82
haploid nucleus		3.0			83
Drosophila melanogaster	2x = 42		39.8(C)	78[f,g]	84, 85, 86
sperm		0.18			87

[a] See footnote [a] for *Table 1*.
[b] See footnote [b] for *Table 1*.
[c] *x* denotes haploid; 2*x*, diploid; 4*x*, tetraploid; 6*x*, hexaploid. Where ?*x* is given, it is uncertain whether the cell is diploid or haploid. The chromosome number is also given.
[d] Satellite DNA present.
[e] Incubation at 50°C.
[f] Reassociation monitored optically by the decrease in hyperchromicity.
[g] Incubation at 66–67°C.

Table 3. DNA and chromosomes of cells: Chordata

Organism	Ploidy[c]	DNA/nucleus (pg)	Mole % G + C[a]	Unique % DNA	Refs
Fish (Teleostei)					
Salmo gairdneri irideus	2x = 58–65				88
erythrocytes		4.9			23
sperm		2.45	43.1(C)		89, 90
Amphibians and reptiles					
Rana clamitans (green frog)				22	91
Rana pipiens	2x = 26		47.3(C)		92, 93
erythrocytes		15.0			94
sperm		6.48			95
Xenopus laevis (African clawed toad)	2x = 36		42.2(C)	54	96, 93, 79, 97
diploid		8.4			98
liver		7.5			99
erythrocytes		6.3			94
Toad					
erythrocytes		7.33			94
sperm		3.70			94
Bufo bufo					
liver		14.2			99
erythrocytes		14.6			100
Uredele, *Axolotl* (*Siredon mexicanum*)					
erythrocyte		6.88			101
sperm		3.36, 4.8			101
Newt					
Triturus cristatus	2x = 24				96
liver		52.5			99
T. vulgaris, liver		71.0			99
Ambystoma mexicanum		77.0			99
Ambystoma tigrinum (tiger salamander)	2x = 28			⩽24	102, 91
Alligator	2x = 32	4.98			103, 23
Birds					
Canary (*Serinus canaria*)	2x = 80				104
liver		4.1			105
Chicken (*Gallus domesticus*)	2x = 78			70	106, 107
erythrocytes		2.34	41(D)		94, 13
sperm		1.26	41(D)		94, 13
Duck	2x = 78				108
erythrocytes		2.65			23
liver		2.1			77
Gallus gallus, sperm		0.73			109

Table 3. Continued

Organism	Ploidy[c]	DNA/nucleus (pg)	Mole % G+C[a]	Unique % DNA	Refs
Sparrow	$2x=76$				110
diploid		1.9			111
Sturnis vulgaris (Starling)					
liver		3.6			105
Mammals					
Boar, sperm		3.44			127
Dog	$2x=78$				123
liver		5.5			77
testes			41.9(C)		128
Guinea-pig	$2x=64$				123
adrenal			42.1(C)		124
liver		5.9			77
Homo sapiens (human)	$2x=46$			64	135, 136
lymphocyte		6.50	40(D)		137[b], 13
sperm		2.44	40(D)		138, 13
Horse	$2x=64$				133
liver		5.8			77
spleen			43(C)		134
Mus musculus (mouse)	$2x=40$			60	115–117
Ehrlich ascites cell		14.0	40(D)		114, 13
fibroblast embryo		5.1	40(D)		112, 13
fibroblast, L-P3			40(D)		13
leucocytes			43.5(C)		118
liver, diploid		6.0			77
lymphocyte		5.31			113[b]
Ox					
liver		6.4	39(D)		125, 13
sperm		3.3	39(D)		125, 13
Pig					
liver		5.0	41(C)		125, 126
spleen			41.2(C)		126
Rabbit					
liver		5.3	39(D)		77, 13
sperm		3.25	39(D)		124, 13
Rattus (rat)	$2x=42$			65	119, 120
lymphocyte		6.05			121
sperm		3.11			121
spleen			43(C)		122
Sheep	$2x=54$				129
leucocyte		6.83			130
sperm		2.9	44(C)		131, 132

[a] See footnote [a] for *Table 1.*
[b] See footnote [b] for *Table 1.*
[c] See footnote [c] for *Table 2.*

Table 4. DNA and chromosomes of cells: animal viruses

Organism	DNA/particle ($\mu g \times 10^{12}$)	Mole % G+C[a]	Unique % DNA	Refs
Adenovirus, 5	120	57.5		139[b], 140
Bovine papilloma (BPV)	8.22	45.5		141[b], 142
Cow pox	27.5	36		143, 144
Equine abortion	417	56.5		137, 145
Herpes simplex	113.6			137
type 1		68		146
Paravirus, H-1 of mice	2.8			147[b]
Parvovirus, HVM of mice	2.8			147[b]
Polyoma virus	5.67	48		148, 149
Pseudorabies	113.6	72		137, 146

[a] See footnote [a] for *Table 1*.
[b] See footnote [b] for *Table 1*.

Table 5. DNA and chromosomes of cells: plants[b]

Organism	Ploidy[c]	DNA/genome (pg)	Mole % G+C[a]	Unique % DNA	Refs
Allium cepa (onion)	$2x=16$	16.8	36.6(C)		150, 151, 152
Arabidopsis thaliana (thale cress)	$2x=10$	0.2			150, 151
Avena sativa (oat)	$6x=42$	13.7			150, 151
Beta vulgaris (beet)	$2x=18$	1.2			150, 151
Brassica napus (turnip)	$2x=38$	1.6			150, 151
Cucurbita melo (melon)	$2x=24$(?)	1.0			150, 151
Glycine max (soybean)	$4x=40$	0.9			150, 151
Gossypium hirsutum (cotton)	$4x=52$	3.0	37.2(C)		150, 151, 153
Hordeum vulgare (barley)	$2x=14$	5.6	43.1(C)		150, 151, 154
Lilium davidii	$2x=24$	43.2			150, 151
Lolium perenne (ryegrass)	$2x=14$	4.9			150, 151
Lycopersicon esculentum (tomato)	$2x=24$	0.75			150, 151
Nicotiana tabacum (tobacco)	$4x=48$	3.9	40.5(C)		150, 151, 154
Oryza sativa (rice)	$2x=24$	1.0			150, 151
Pisum sativum (pea)	$2x=14$	4.9	41.9(C)		150, 151, 154
Raphanus sativus (radish)	$2x=18$	0.4			150, 151
Secale cereale (rye)	$2x=14$	9.5			150, 151

Table 5. Continued

Organism	Ploidy[c]	DNA/genome (pg)	Mole % G + C[a]	Unique % DNA	Refs
Solanum tuberosum (potato)	$4x=48$	2.1			150, 151
Spinacia oleracea (spinach)	$2x=12$	1.0	37.4(C)		150, 151, 155
Triticum aestivum (hexaploid wheat)	$6x=42$	17.3		20	150, 151, 156
Vicia faba (broad bean)	$2x=12$	13.3		⩽15	150, 151, 157
Zea mays (maize)	$2x=20$	3.9	46(C)		150, 151, 158

[a] See footnote [a] for *Table 1*.
[b] The data in this table has been kept to a minimum, further details can be obtained from *Plant Molecular Biology LabFax*.
[c] See footnote [c] for *Table 2*.

DNA/Chromosomes

REFERENCES

1. Marmur, J. and Doty, P. (1962) *J. Mol. Biol.*, **5**, 109.

2. De Ley, J. and Schell, J. (1963) *J. Gen. Microbiol.*, **33**, 243.

3. Okanishi, M., Akagawa, H. and Umezawa, H. (1972) *J. Gen. Microbiol.*, **72**, 49.

4. Roberstad, Hester and Gordon, cited in Normore, W.M. (1976) *Handbook of Biochemistry & Molecular Biology (3rd edn)* (G. D. Fasman ed.). CRC Press, Ohio, Vol. II, p. 65.

5. De Ley, J., Bernaerts, M., Rassel, A. and Guilmot, J. (1966) *J. Gen. Microbiol.*, **43**, 7.

6. Mandel, M., cited in De Ley, J. (1970) *J. Bacteriol.*, **101**, 738.

7. De Ley, J., Kersters, K., Khan-Matsubara, J. and Shewan, J.M. (1970)*Antonie van Leeuwenhoek J. Microbiol. Serol.*, **36**, 193.

8. Colwell, R. and Mandel, M. (1964) *J. Bacteriol.*, **87**, 1412.

9. Skyring, G.W. and Quadling, C. (1970)*Can. J. Microbiol.*, **16**, 95.

10. Jones and Bradley (1964) *Dev. Ind. Microbiol.*, **5**, 267.

11. De Ley, J. and Park, I.W. (1966) *Antonie van Leeuwenhoek J. Microbiol. Serol.*, **32**, 6

12. Fitz-James, P.C. and Young, I.E. (1959) *J. Bacteriol.*, **78**, 743.

13. Szybalski, W. (1968) *Methods Enzymol.*, **12B**, 330.

14. Hodson, P.H. and Beck, J.V. (1960) *J. Bacteriol.*, **79**, 661.

15. Dubnau, D., Smith, I., Morell, P. and Marmur, J. (1965) *Proc. Natl. Acad. Sci. USA*, **54**, 491.

16. Spiegelman, S., Aronson, A. and Fitz-James, P.C. (1958) *J. Bacteriol.*, **75**, 102.

17. Ikeda, Y., Saito, H., Miura, K.I., Takagi, J. and Aoki, H. (1965) *J. Gen. Appl. Microbiol.*, **11**, 181.

18. Welker, N.E. and Campbell, L.L. (1967) *J. Bacteriol.*, **94**, 1124.

19. Aubert, J.P., Ryter, A. and Schaeffer, P. (1968) *Ann. Inst. Pasteur*, **115**, 989.

20. Tonomura, B., Malkin, R. and Rabinowitz, J.C. (1965) *J. Bacteriol.*, **89**, 1438.

21. Schildkraut, C.L., Marmur, J. and Doty, P. (1962) *J. Mol. Biol.*, **4**, 430.

22. Saunders, G.F., Campbell, L.L. and Postgate, J.R. (1964) *J. Bacteriol.*, **87**, 1073.

23. Vendrely, R. (1958) *Ann. Inst. Pasteur*, **94**, 142.

24. De Ley, J. (1970) *J. Bacteriol.*, **101**, 738.

25. Gillies, N.E. and Alper, T. (1960) *Biochim. Biophys. Acta*, **43**, 182.

26. Mitchell, Hendrie, Margaret and Shewan (1969) *J. Appl. Microbiol.*, **32**, 40.

27. Marmur, J., Falkow, M. and Mandel, S. (1963) *Ann. Rev. Microbiol.*, **17**, 329.

28. Ouellette, C.A., Burris, R.H. and Wilson, P.W. (1969) *Antonie van Leeuwenhoek J. Microbiol. Serol.*, **35**, 275.

29. Gasser, F. and Mandel, M. (1968) *J. Bacteriol.*, **96**, 580.

30. Cantoni, Manachini and Craveri (1965) *Ann. Microbiol. Enzymol.*, **15**, 151.

31. Jackson, J.F., Moriarty, D.J.W. and Nicholas, D.J.D. (1968) *J. Gen. Microbiol.*, **53**, 53.

32. Anderson, J.R., Pramer, D. and Davis, F.F. (1965) *Biochim. Biophys. Acta*, **108**, 155.

33. Mandel, M. (1966) *J. Gen. Microbiol.*, **43**, 273.

34. Colwell, R.R., Citarella, R.V. and Ryman, I. (1965) *J. Bacteriol.*, **90**, 1148.

35. Colwell, R.R. and Mandell, M. (1965) *J. Bacteriol.*, **89**, 454.

36. Heberlein, G.T., De Ley, J. and Tijtgat, R. (1967) *J. Bacteriol.*, **94**, 116.

37. Lark, K.G. and Maaløe, O. (1956) *Biochim. Biophys. Acta*, **21**, 448.

38. Baptist, J.N., Shaw, C.R. and Mandel, M. (1969) *J. Bacteriol.*, **99**, 180.

39. Belozerskii and Spirin (1960) in *The Nucleic Acids*, (Chargaff and Davidson, eds). Academic Press, New York, Vol. 3, p. 147.

40. Jones and Walker (1968) *Arch. Biochem. Biophys.*, **128**, 597.

41. Ottogalli and Galli (1967) *Ann. Microbiol. Enzymol.*, **17**, 199.

42. Frontali, C., Hill, L.R. and Silvestri, L.G. (1965) *J. Gen. Microbiol.*, **38**, 243.

43. Monson, Bradley, Enquist and Cruces (1969) *J. Bacteriol.*, **99**, 702.

44. Edelman, M., Swinton, D., Schiff, J.A., Epstein, H.T. and Zeldin, B. (1967) *Bacteriol. Rev.*, **31**, 315.

45. Friedman, S. and De Ley, J. (1965) *J. Bacteriol.*, **89**, 95.

46. Leedale, G. F. (1958). *Nature*, **181**, 502.

47. Brawerman, G., Rebman, C.A. and Chargaff, E. (1960) *Nature*, **187**, 1037.

48. Edelman, M., Schiff, J.A. and Epstein, H.T. (1965) *J. Mol. Biol.*, **11**, 769.

49. Suyama, Y. and Preer, J.R. (1965) *Genetics*, **52**, 1051.

50. Allen, S. and Gibson, I. (1972) *Biochem. Genet.*, **6**, 293.

51. Gintsburg, G.I. (1961) *Zh. Obshch. Biol.*, **22**, 452.

52. Gintsburg, G.I. (1963), cited in Antonov (1965) *Usp. Sovrem. Biol.*, **60**, 161.

53. Scherbaum (1957) *Exp. Cell Res.*, **13**, 24.

54. Swartz, M.N., Trautner, T.A. and Kornberg, A. (1962) *J. Biol. Chem.*, **237**, 1961.

55. Mandel, M. and Honigberg, B.M. (1964) *J. Protozool.*, **11**, 114.

56. Roskin, G. and Schischliaiewa, S. (1928) *Arch. Protistenk.*, **60**, 460.

57. Riou, G., Pautrizel, R. and Paoletti, C. (1966) *C. R. Acad. Sci. (Paris)*, **262**, 2367

58. Craig, Leach and Carr (1969) *Arch. Microbiol.*, **65**, 218.

59. Chun, E.H.L., Vaughan, M.H. and Rich, A. (1963) *J. Mol. Biol.*, **7**, 130.

60. Wells, R. and Sager, R. (1971) *J. Mol. Biol.*, **58**, 611.

61. Elliott, C. G. (1960) *Genet. Res.*, **1**, 462.

62. Dutta, S.K., Richman, N., Woodward, V.W. and Mandel, M. (1967) *Genetics*, **57**, 719.

63. Heagy, F.C. and Roper, J.A. (1952) *Nature*, **170**, 713.

64. Nakase and Komataga (1968) *J. Gen. Appl. Microbiol.*, **14**, 345.

65. Firtel, R.A. and Bonner, J. (1972) *J. Mol. Biol.*, **66**, 339.

66. Fincham, J.R.S. and Day, P.R. (1963) *Fungal Genetics*. F.A. Davis, Philadelphia.

67. Luck, D.J.L. and Reich, E. (1964) *Proc. Natl. Acad. Sci. USA* ., **52**, 931.

68. Brooks and Huang (1972) *Biochem. Genet.*, **6**, 41.

69. Guttes, E. (1971) Unpublished. University of Texas, Dallas.

70. Mandel, M. (1970) in *A Handbook of Biochemistry (2nd edn)* (Sober, ed.). Chemical Rubber Co., Cleveland, H-75.

71. Fouquet, H., Bierweiler, B. and Sauer, H.W. (1974) *Eur. J. Biochem.*, **44**, 407.

72. Hawthorne, D. C. and Mortimer, R.K. (1968). *Genetics*, **60**, 735.

73. Williamson, D. and Scopes, A.W. (1961) *Exp. Cell Res.*, **24**, 151.

74. Storck, R. (1966) *J. Bacteriol.*, **91**, 227.

75. Ogur, Minckler, Lindegren and Lindegren (1952) *Arch. Biochem. Biophys.*, **40**, 175.

76. Mirsky, A.E. and Ris, H. (1951) *J. Gen. Physiol.*, **31**, 451.

77. Vendrely, R. and Vendrely, C. (1949) *Experientia*, **5**, 327.

78. Chargaff, Lipshitz and Green (1952) *J. Biol. Chem.*, **195**, 155.

79. Davidson, E.H. and Britten, R.J. (1973) *Q. Rev. Biol.*, **48**, 565.

80. Britten, R.J., Graham and Henrey (1972) *Carnegie Inst. Wash. Year Book*, **71**, 270

81. Jost, E. and Mameli, M. (1972) *Chromosoma*, **37**, 201.

82. Artom, C. (1928) *C. R. Sol. Biol.*, **99**, 29.

83. Rheinsmith, E.L., Hinegardner, R. and Bachmann, K. (1974) *Comp. Biochem. Physiol.*, **48B**, 343.

84. Guyénot, E. and Naville, A. (1929) *Cellule*, **39**, 25.

85. Argyrakis, M.P. and Bessman, M.J. (1963) *Biochim. Biophys. Acta*, **72**, 122.

86. Wu, J., Hurn, J. and Bonner, J. (1972) *J. Mol. Biol.*, **64**, 211.

87. Rasch, Barr and Rasch (1971) *Chromosoma*, **33**, 1.

88. Ohno. S., Stenius, C., Faisst, E. and Zenzes, M.T. (1965) *Cytogenetics*, **4**, 117.

89. Vendrely, R. and Vandrely, C. (1953) *Nature*, **172**, 30.

90. Felix, K., Jilke, I. and Zahn, R.K. (1956) *Hoppe-Seyler's Z. Physiol. Chem.*, **303**, 140.

91. Straus, N.A. (1971) *Proc. Natl. Acad. Sci. USA* , **68**, 799.

92. Porter, K. R. (1941) *Biol. Bull.*, **80**, 238.

93. Dawid, I.B. (1965) *J. Mol. Biol.*, **12**, 581.

94. Mirsky, A.E. and Ris, H. (1949) *Nature*, **163**, 666.

95. England, M.C. and Mayer, D.T. (1957) *Exp. Cell Res.*, **12**, 249.

96. Mikamo, K. and Witschi, E. (1966) *Cytogenetics*, **5**, 1.

97. Davidson, E.H., Hough, B.R., Amenson, C.S. and Britten, R.J. (1973) *J. Mol. Biol.*, **77**, 1.

98. Birstow and Deuchar (1964) *Exp. Cell Res.*, **35**, 580.

99. Conger, A.D. and Clinton, J.H. (1973) *Radiat. Res.*, **54**, 69.

100. Bachmann, K. (1970) *Chromosoma*, **29**, 365.

101. Edström, J.E. (1964) *Biochim. Biophys. Acta*, **80**, 399.

102. Carrick, R. (1934) *Trans. Roy. Soc. Edinburgh*, **58**, 63.

103. Cohen, M. M. and Gans, C. (1970) *Cytogenetics*, **9**, 81.

104. Ohno. S., Stenius, C., Christia, L.C., Becak, W. and Becak, M.L. (1964) *Chromosoma*, **15**, 280.

105. Bachmann, K., Harrington, B.A. and Craig, J.P. (1972) *Chromosoma*, **37**, 405.

106. Owen, J.T.T. (1965) *Chromosoma*, **16**, 601.

107. Sanchez de Jimenéz, E., González, J.L., Domínguez, J.L. and Saloma, E.S. (1974). *Eur. J. Biochem.*, **45**, 25.

108. Hammer, B. (1966) *Hereditas*, **55**, 367.

109. Eapen and Raza Nasir (1963) *Indian Vet. J.*, **40**, 803.

110. Brink, J.M. van (1959) *Chromosoma*, **10**, 1.

111. Vendrely, R. and Vendrely, C. (1950) *C. R. Acad. Sci. Ser. D*, **230**, 788.

112. Bassleer, R. (1964) *C. R. Soc. Biol.*, **158**, 384.

113. Menton, Willms and Wright (1953) *Cancer Res.*, **13**, 729.

114. Leuchtenberger, Klein and Klein (1952) *Cancer Res.*, **12**, 480.

115. Makino, S. (1941) *J. Fac. Sci. Hokkaido Imp. Univ.*, VI, **7**, 305.

116. Straus, N.A. (1976) in *Handbook of Biochemistry & Molecular Biology (3rd edn)* (G. D. Fasman ed.). CRC Press, Ohio, Vol. II, p. 319.

117. Straus, N.A. and Birboim, H.C. (1974) *Proc. Natl. Acad. Sci. USA* ., **71**, 2992.

118. Penn, N.W., Suwalski, R., O'Riley, C., Bojanowski, K. and Yura, R. (1972) *Biochem. J.*, **126**, 781

119. Makino, S. and Asana, J.J. (1948) *Chromosoma*, **3**, 208.

120. Holmes, D.S. and Bonner, J. (1974) *Biochemistry*, **13**, 841.

121. Sandritter, Müller and Gensecke (1960) *Acta Histochemica*, **10**, 139.

122. Kleinschmidt and Manthey (1958) *Arch. Biochem. Biophys.*, **73**, 52.

123. Awa, A. *et al.* (1959) *Jap. J. Zool.*, **12**, 257.

124. Bransome, Jr., E.D. and Chargaff, E. (1964) *Biochim. Biophys. Acta.*, **91**, 180.

125. Vendrely, R. and Vendrely, C. (1948) *Experienta*, **4**, 434.

126. Chargaff, E. and Lipshitz (1953) *J. Am. Chem. Soc.*, **75**, 3658.

127. Ivanov, Korban and Sharobaiko (1969) *Bull. Eksper. Biol. Med.*, **67**, 46.

128. Busch, E.W., von Borcke, I.M., Greve, H. and Thorn, W. (1968) *Hoppe-Seyler's Z. Physiol. Chem.*, **349**, 801.

129. Melander, Y. (1959) *Hereditas*, **45**, 649.

130. Mandel, M., Metais and Cuny (1950) *C. R. Acad. Sci.*, **231**, 1172.

131. Aberg, B. and Gillner, M. (1966) *Acta Physiol. Scand.*, **66**, 106.

132. Wyatt, G.R. (1951) *Biochem. J.*, **48**, 584.

133. Sasaki, M. S. and Makino, S. (1962) *J. Hered.*, **53**, 157.

134. Daly, Allfrey and Mirsky (1950) *J. Gen. Physiol.*, **33**, 497.

135. Tijio, J. H. and Levan, A. (1956) *Hereditas*, **42**, 1.

136. Saunders, G.F., Shirakawa, S., Saunders, P.P., Arrighi, F.E. and Hsu, T. (1972) *J. Mol. Biol.*, **63**, 323.

137. Darlington, R.W. and Randall, C.C. (1963) *Virology*, **19**, 322.

138. Leuchtenberger, Leuchtenberger and Davis (1954) *Am. Pathol.*, **30**, 65.

139. Allison, A.C. and Burke, D.C. (1962) *J. Gen. Microbiol.*, **27**, 181.

140. Piña, M. and Green, M. (1965) *Proc. Natl. Acad. Sci. USA* , **54**, 547.

141. Lang, D., Bujard, H., Wolff, B. and Russell, D. (1967) *J. Mol. Biol.*, **23**, 163.

142. Thomas, C.A. and MacHattie, L.A. (1967) *Ann. Rev. Biochem.*, **36**, 485.

143. Joklik, W.K., cited in Shapiro, H.S. (1976) in *Handbook of Biochemistry & Molecular Biology (3rd edn)* (G. D. Fasman ed.). CRC Press, Ohio, Vol. II, p. 310.

144. Joklik, W.K. (1962) *J. Mol. Biol.*, **5**, 265.

145. Soehner, R.L., Gentry, G.A. and Randall, C.C. (1965) *Virology*, **26**, 394.

146. Plummer, G., Goodheart, C.R., Henson, D. and Bowling, C.P. (1969) *Virology*, **39**, 134.

147. McGeoch, D.J., Crawford, L.V. and Follett, E.A.C. (1970) *J. Gen. Virol.*, **6**, 33.

148. Crawford, L.V. (1974) *Virology*, **22**, 149.

149. Yamagishi, H., Yoshizako, F. and Sato, K. (1966) *Virology*, **30**, 29.

150. Bennett *et al.* (1982) *Proc. R. Soc. London, Series B*, **216**, 179.

151. Bennett and Smith (1976) *Phil. Trans. R. Soc. London, Series B*, **274**, 227.

152. Uryson, S.O. and Belozerskii, A.N. (1959) *Dokl. Akad. Nauk SSSR*, **125**, 1144.

153. Sulimova, G.E., Mazin, A.L., Vanyushin, B.F. and Belozerskii, A.N. (1970) *Dokl. Akad. Nauk SSSR*, **193**, 1422.

154. Vanyushin, B.F., Kadyrova, D.Kh., Karimov, Kh.Kh. and Belozerskii, A.N. (1971) *Biokhimya*, **36**, 1251.

155. Bard, S.A. and Gordon, M.P. (1969) *Plant Physiol.*, **44**, 377.

156. Bendich, A.J. and McCarthy, B.J. (1970) *Genetics*, **65**, 545.

157. Straus (1972) *Carnegie Inst. Wash. Year Book*, **71**, 257.

158. Ergle, D.R. and Katterman, F.R. (1961) *Plant Physiol.*, **36**, 811.

CHAPTER 6
THE COMPOSITION AND STRUCTURE OF MEMBRANES

J. M. Graham

1. GENERAL

1.1. Lipids

All membranes of eucaryotic organisms and the cytoplasmic membranes of procaryotes contain a bilayer of amphipathic lipids. Even the outer membrane of Gram-negative bacteria can be regarded as having this structure; the inner half of the bilayer being composed of lipid molecules, the outer half consisting of the lipid chains of lipopolysaccharide.

The lipid bilayer is permeable to small non-ionized molecules (e.g. glycerol, urea, water, oxygen and carbon dioxide) and to lipophilic molecules, such as fatty acids. Lipids also provide the proper environment for the functioning of many membrane proteins.

1.2. Proteins

All membranes contain proteins which either (a) span the bilayer (integral or transmembrane proteins), (b) are partially buried in one half of the bilayer, or (c) are bound to the bilayer surface or to type (a) or (b) protein.

Proteins provide principally:
(i) the selective ionic permeability of the membrane and hence its electrical properties;
(ii) the means of energy transduction;
(iii) the means of responding to a signal on one side and of propagating a response on the other side;
(iv) transport systems for a range of hydrophilic metabolites (e.g. glucose and amino acids);
(v) structural functions through their interactions with non-membrane macromolecules.

1.3. Carbohydrates

Most membranes contain glycosylated molecules, the carbohydrate is always bound covalently to another membrane component. Glycoproteins and glycolipids contain short oligosaccharide chains (often branched). Proteoglycans that occur in eucaryotes are predominantly carbohydrate, consisting of long (largely unbranched) chains of sugar derivatives linked to relatively short polypeptide segments. Lipopolysaccharides are found only in the outer membrane of Gram-negative bacteria; like proteoglycans they have long chains of complex saccharides.

In eucaryotes, glycolipids and glycoproteins are highly site-specific; their major roles being as surface receptors and antigens. Proteoglycans are also confined to the surface. Frequently their carbohydrate (called glycosaminoglycan) chains exist independently of any protein connection to the membrane; they are involved in cell–cell adhesions, communication between cells and provide an intercellular matrix.

2. LIPIDS

2.1. General lipid composition

In eucaryotes the protein concentration in membranes of cytoplasmic organelles tends to be higher than that of the surface membrane, and the procaryotic cytoplasmic membrane resembles eucaryotic organelle membranes in its protein content. Consequently, the surface membrane of eucaryotes tends to have the lowest lipid:protein ratio of all membranes (*Table 1*).

Table 1. Protein:lipid ratios in membranes

Membrane[a]	Protein:lipid	Membrane[a]	Protein:lipid
Animal			
Myelin	0.25	Erythrocyte	1.1
Surface (liver)	1.5	Nuclear (liver)	2.0
Rough ER (liver)	2.5	Smooth ER (liver)	2.1
Inner mit. (liver)	3.6	Outer mit. (liver)	1.2
Golgi (liver)	2.4	Sarcoplasmic reticulum	3.0
Retinal rod	1.0		
Plant			
Surface	0.9	Chloroplast	1.9
Saccharomyces			
Surface	1.2		
Procaryote surface			
Bacillus	2.8	*Micrococcus*	2.4
Staphylococcus	2.4	*Escherichia coli*	2.8
Outer membrane of Gram-negative bacteria	2.2		

[a] Abbreviations: ER, endoplasmic reticulum; mit., mitochondrial.

All membranes contain phospholipids, which are normally the major amphipathic lipids present (*Table 2*). Glycolipids and sterols are usually concentrated in the plasma membrane of mammalian cells, while in plants only sterols show this preferential localization; the glycolipids predominate in the chloroplast. Bacterial membranes may also be rich in glycolipid but, with the exception of mycoplasmas, they lack any sterol.

For general information on lipid components of animal cells see refs 1–3; more specifically, for plant cell membranes see ref. 4, and for bacterial membranes see ref. 5.

2.2. Phospholipids

Derivatives of diacylglycerol

All phospholipids in procaryotes and plant cells and most of the phospholipids of animal cells are derivatives of diacylglycerol-3-phosphate (phosphatide).

Table 2. General lipid classes (% of total lipid by weight)

Membrane	Phospholipid	Glycolipid	Sterol
Mammalian			
Plasma	50–60	5–17[a]	15–22
Endoplasmic reticulum	70–80	<5	5–10
Mitochondria (inner)	80–90	<5	<5
Mitochondria (outer)	80–90	<5	5–8
Lysosomes	70–80	5–10	10–15
Nuclear	85–90	<5	10–15[b]
Golgi	85–90	<5	5–10
Peroxisomes	90–95	<5	<5
Myelin	50–60	15–25	20–25
Erythrocyte	70–80	5–10	20–25
Plant			
Surface	30–65	10–20	25–50
Mitochondria	90–95	<5	<5
Chloroplast envelope	20–30	65–80	<5
Chloroplast lamellae	35–45	50–70	<5
Endoplasmic reticulum	70–80	5–15	10–20
Bacterial			
Cytoplasmic	50–90	10–50	None

[a] Glycolipids in mammalian cells are rarely reported as a percentage of the total lipid. The glycolipid in the plasma membrane of cultured cells has been given (17) as between 10 and 30% of the 'total polar lipids'.

[b] A variable proportion of nuclear membrane sterol is reported to be as the ester rather than the free sterol.

In eucaryotes the two variable acyl residues are derived from unbranched fatty acids (see *Table 3*) of between C14 and C24 (animals) and C16 and C18 (plants). They are frequently unsaturated or polyunsaturated: in animals up to six *cis*-double bonds may occur, in plants rarely more than three. In animals one of the fatty acids is normally saturated and the other unsaturated.

In procaryotes, C15 to C19 acids are the most common (unlike eucaryotes they are not synthesized by the stepwise addition of 2C units, so both odd and even carbon numbers are equally possible). Branched or cyclopropane derivatives may occur (see *Figure 1*). Polyunsaturated acids are absent from procaryotes except for *Mycobacterium phlei* (5).

In eucaryotes the phosphate group is commonly linked to one of the following (see *Figure 2*): a nitrogenous base (choline, ethanolamine or serine); inositol; or glycerol. Phosphatidylglycerol (PG) is uncommon in animals but its derivative diphosphatidylglycerol (DPG), or cardiolipin, is common in both eucaryotes and procaryotes.

In bacteria, PG and phosphatidylethanolamine (PE) are common and derivatives of both of these phospholipids are encountered. The CH_2OH group of the glycerol may be linked through an acyl group to an amino acid (amino-acyl ester of PG) and the amine group of PE may be methylated. Mannosyl derivatives of phosphatidylinositol (PI) may also be present.

Table 3. Common fatty acids in eucaryotes

Carbon number	Number of double bonds	Name	Formula
12	0	Laurate	$CH_3-(CH_2)_{10}-COO^-$
14	0	Myristate	$CH_3-(CH_2)_{12}-COO^-$
16	0	Palmitate	$CH_3-(CH_2)_{14}-COO^-$
16	1	Palmitoleate	$CH_3-(CH_2)_5-CH=CH-(CH_2)_7-COO^-$
18	0	Stearate	$CH_3-(CH_2)_{16}-COO^-$
18	1	Oleate	$CH_3-(CH_2)_7-CH=CH-(CH_2)_7-COO^-$
18	2	Linoleate	$CH_3-(CH_2)_4-(CH=CH-CH_2)_2-(CH_2)_6-COO^-$
18	3	Linolenate	$CH_3-CH_2-(CH=CH-CH_2)_3-(CH_2)_6-COO^-$
20	0	Arachidate	$CH_3-(CH_2)_{18}-COO^-$
20	4	Arachidonate	$CH_3-(CH_2)_4-(CH=CH-CH_2)_4-(CH_2)_2-COO^-$
22	0	Behenate	$CH_3-(CH_2)_{20}-COO^-$
24	0	Lignocerate	$CH_3-(CH_2)_{22}-COO^-$

Figure 1. Fatty acids of bacterial membranes.

Branched chain 'iso' fatty acids

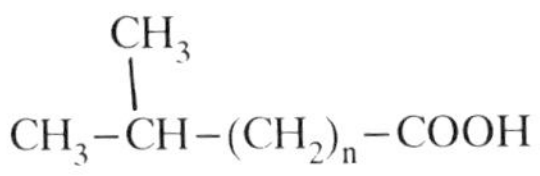

$$\begin{array}{l} \quad\;\; CH_3 \\ \quad\;\;\; | \\ CH_3-CH-(CH_2)_n-COOH \end{array}$$

Branched chain 'anteiso' fatty acids

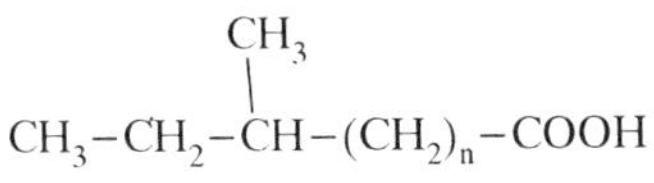

$$\begin{array}{l} \qquad\qquad\;\; CH_3 \\ \qquad\qquad\;\;\; | \\ CH_3-CH_2-CH-(CH_2)_n-COOH \end{array}$$

Cyclopropane fatty acids

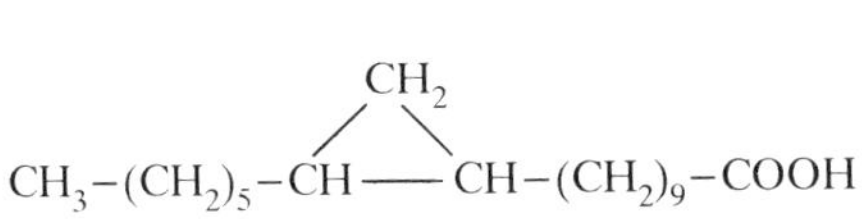

$$\begin{array}{c} CH_2 \\ / \;\; \backslash \\ CH_3-(CH_2)_5-CH \text{---} CH-(CH_2)_9-COOH \end{array}$$

Figure 2. Common phospholipids: molecular structure. A, phosphatidylcholine; B, phosphatidylethanolamine; C, phosphatidylserine; D, phosphatidylinositol; E, phosphatidylglycerol; F, diphosphatidylglycerol (cardiolipin); G, sphingomyelin. In G the structure within the dotted box is sphingosine and that within the continuous box is ceramide.

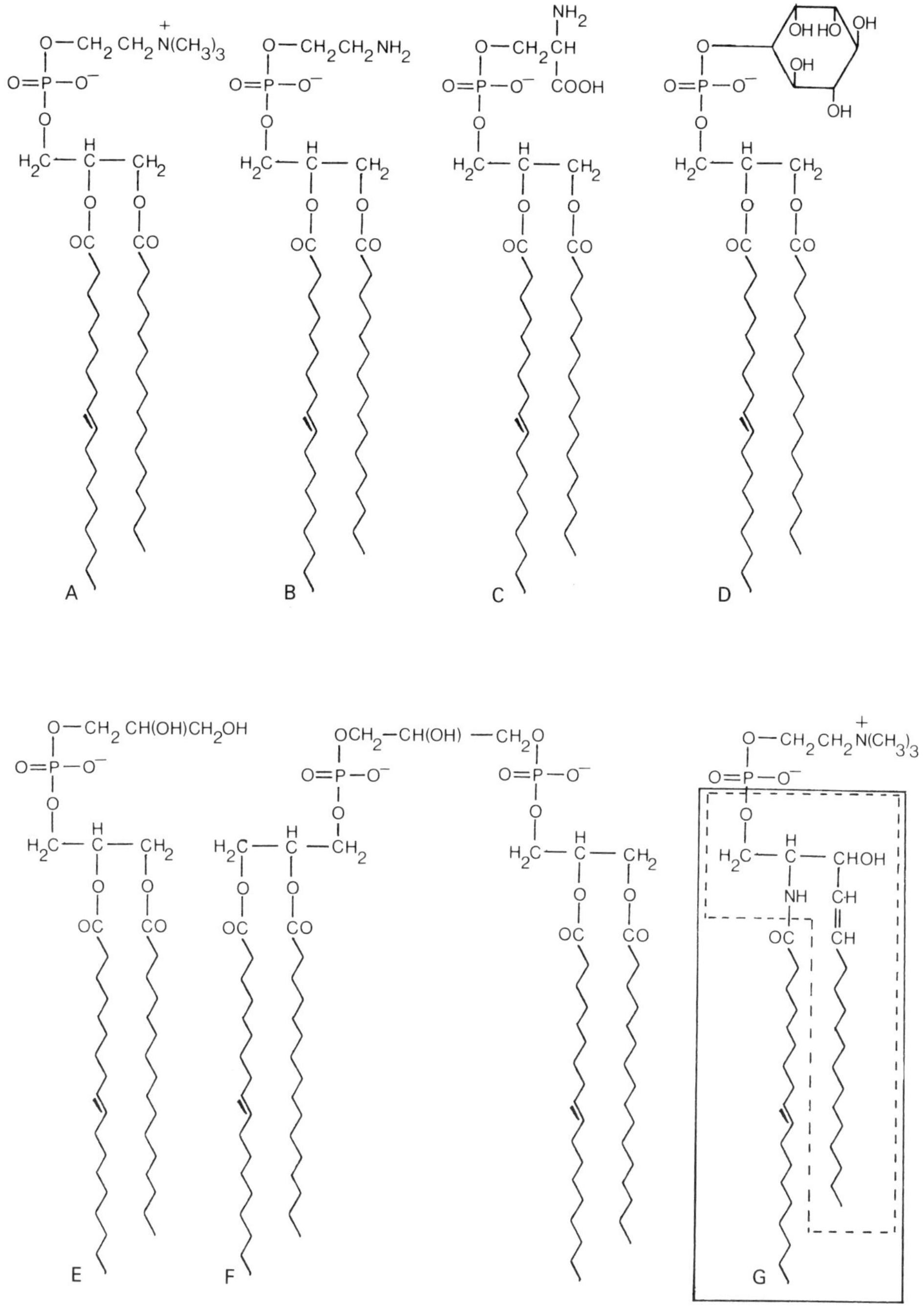

Sphingomyelin
Sphingomyelin (SM) is unique to animals. The sphingosine molecule (*Figure 2G*), which is equivalent to 1-acylglycerol of other phospholipids, can only have one variable acyl group, which is linked to the free amine group (amide-linked). The terminal OH group of the acylsphingosine (ceramide) is linked to phosphoryl choline. SM and phosphatidylcholine (PC) are sometimes grouped together as the choline-containing phospholipids in animals.

Lysophospholipids
In animals, small amounts (1–3% of total lipid) of lysophosphatidylcholine and lysophosphatidyl ethanolamine are present (see *Table 4*). They are deficient in one acyl group and are probably derived from the parent molecule by phospholipase action. They are membrane perturbants and can act as detergents.

Table 4. Occurrence of lysophospholipids in animal cell membranes

Membrane type	**Phospholipid (as a % of total phospholipid)**	
	Lyso-PC[a]	**Lyso-PE**[b]
Plasma membrane	2–4	<1
Endoplasmic reticulum	3–7	<1
Golgi	3–7	3–7
Lysosomes	2–4	<1

[a] Lyso-PC = lysophosphatidylcholine.
[b] Lyso-PE = lysophosphatidyl ethanolamine.

Ether lipids (plasmalogens)
The hydrocarbon chain on the C1 position of the glycerol backbone is linked via an ether group rather than an ester (8). There is always an ethylenic link adjacent to the ether group. They are found in both eucaryotes (mainly animals) and procaryotes, normally at low levels (1–5% of the total lipid). However, in specific sites in certain animal tissues and, in particular, micro-organisms they can be major components (5, 8).

In eucaryotes ether-linked derivatives of PE and PC are known; in procaryotes derivatives of PE and phosphatidylserine (PS) have been reported (see *Table 5*).

Function of phospholipids
The specific function of phospholipids, other than as the essential components of the lipid bilayer, is not entirely clear. Phosphatidylinositol (PI) is known to be involved in intracellular signalling (9), and other phospholipids are thought to be specifically involved in the modulation of the activity of certain membrane proteins (10, 11).

Phospholipid profiles of membranes (4–7, 12–14)
The relative amounts of different common phospholipids are given for the membranes of animal cells (*Table 6*), plant cells (*Table 7*) and microbial cells (*Table 8*). See ref. 6 for a more detailed analysis of phospholipids in membranes.

Table 5. Occurrence of plasmalogens in membranes

Membrane type	Phospholipid (as % of total phospholipid)		
	Choline plas[a]	Ethan plas[b]	Serine plas[c]
Gram-negative cocci cytoplasmic[d]	—	30–54	3–25
Sarcoplasmic reticulum	5–10[e]	60–70[e]	

[a] Choline plasmalogen.
[b] Ethanolamine plasmalogen.
[c] Serine plasmalogen.
[d] Of the remaining phospholipid, 7–44% is PE and 10–20% is PS.
[e] These figures are the percentages of ether-linked choline-containing and ethanolamine-containing phospholipid. About 50% of total phospholipid is ether lipid.

Table 6. Phospholipid profile of animal (rat) cell membranes[a]

Membrane[c]	Phospholipid (% of total phospholipid)[b]						
	PC	PE	PI	PS	PG	DPG	SM
Plasma	35–45	17–22	6–8	5–10	2–5	<1	15–18
ER	50–60	17–22	8–10	5–10	<1	<1	3–5
Mit. inner	43–48	23–28	5–12	1–2	2–3	16–20	1–2
Mit. outer	50–55	23–25	12–15	2–3	2–3	2–3	3–5
Lysosome	25–40	13–18	8–10	3–8	<1	5–8	20–35
Nuclear	50–60	15–23	5–8	3–5	<1	<1	2–3
Golgi	45–55	15–20	6–9	3–5	<1	<1	6–13
Peroxisome	70–75	15–20	2–5	2–5	<1	<1	<1
Myelin	30–40	35–45	3–5	10–15	<1	<1	15–20

[a] All membranes except myelin are from liver.
[b] <1 means that amounts are beyond the lower limit of detection.
[c] Abbreviations: ER, endoplasmic reticulum; Mit., mitochondrial.

Table 7. Phospholipid profiles of plant cell membranes

Membrane[b]	Phospholipid (as % of total phospholipid)[a]					
	PC	PE	PI	PS	PG	DPG
Plasma	30–35	40–50	15–20	<1	<1	3–5
ER	35–60	15–30	10–20	1–2	2–4	2–10
Mit. inner	25–35	30–35	3–8	5–25	2–5	15–20
Mit. outer	40–55	23–28	5–20	8–12	8–12	3–12
Chlor. envelope[c]	65–75	<1	3–5	<1	20–35	<1
Chlor. lamellae[c]	25–35	<1	<1	<1	60–75	<1

[a] <1 means that amounts are beyond the lower limit of detection.
[b] Abbreviations: ER, endoplasmic reticulum; Mit., mitochondrial; Chlor., chloroplast.
[c] Because the glycolipid content of chloroplast membranes (envelope and lamellae) is over 60%, the amount of each individual phospholipid as a % of total lipid is much lower than the figures given in the table. In all other membranes, phospholipid is the major lipid type.

Table 8. Phospholipid profile of bacterial cytoplasmic membranes[a]

Organism	Phospholipid (% of total phospholipid)[b]					
	PC	PE	NMePE[c]	PG	OAAPG[d]	DPG
Gram-positive						
Micrococcus	<1	<1	<1	60–70	16–18	4–20
Bacillus	<1	20–45	<1	25–45	5–15	10–50
Gram-negative						
Azotobacter	1–2	60–70	3–8	25–30	<1	2–5
Rhodopseudomonas	10–15	45–50	<1	40–45	<1	<1

[a] The values for phospholipid composition vary very widely: average values would be of little use; they depend not only on the type of organism but also on the growth conditions. Generally, PC is absent from Gram-positive bacteria. The outer membranes of Gram-negative bacteria generally contain PE, PG and DPG.
[b] <1 means below the lower limits of detection.
[c] NMePE = *N*-methylated PE.
[d] OAAPG = *O*-aminoacyl ester derivative of PG.

Fatty acids in phospholipids (6, 7, 12)
The range of fatty acid residues present in phospholipid molecules from animal membranes shows some degree of specificity for phospholipid type, but, with the exception of those present in the plasma membrane, show rather little membrane specificity within a particular cell type (see *Tables 9* and *10*). The longest acyl chains (C24) are confined to sphingomyelin and are therefore very prominent in myelin.

Table 9. Fatty acid profiles of phospholipids from intracellular membranes (rat liver)

Fatty acid	% of total			
	PC	PE	PI	DPG
16:0	22–30	22–30	16–28	7–10
16:1	3–4	3–4	5–6	7–10
18:0	22–26	22–28	35–45	3–5
18:1	12–16	8–12	10–18	18–20
18:2	12–16	5–10	2–8	55–60
20:0	—	—	—	—
20:3	1–2	—	1–2	1–2
20:4	15–20	15–24	18–24	—
22:6	2–3	4–12	1–4	—
24:0	—	—	—	—

Table 10. Fatty acid profiles of phospholipids of the plasma membrane[a]

Fatty acid	% of total				
	PC	PE	PI	PS	SM
16:0	32(37)	30(23)	30	(8)	36(35)
16:1	3(2)	1(2)	8	(2)	4(2)
18:0	35(13)	31(13)	36	(37)	25(13)
18:1	10(23)	10(15)	13	(14)	2(3)

Table 10. Continued

Fatty acid	% of total				
	PC	PE	PI	PS	SM
18:2	8(16)	6(7)	2	(3)	1(3)
20:0	—	—	—	—	—
20:3	—	1(1)	—	—	—
20:4	8(8)	16(23)	8	(23)	18(—)
22:6	1(—)	3(2)	—	(2)	—
24:0	—	—(2)	—	(2)	—(20)
24:1	—	—(2)	—	(2)	—(15)

[a] The figures in brackets are the fatty acid profiles for phospholipids from the human erythrocyte membrane. Figures not in brackets are for the rat liver plasma membrane.

In plants, on the other hand, (12) the fatty acid components of all phospholipids appear to be specific to the location within the cell, and they are significantly different to those present in animal phospholipids (*Table 11*).

In both animals and plants there is considerable species variability in the amount of each type of fatty acid. *Table 12* gives the amounts of the major fatty acids in the total phospholipids from a number of species. This species variability is also probably the cause of the large range of concentrations given in *Table 11*.

Table 11. Fatty acid profile of plant cell phospholipids (total)

Membrane[a]	% of total						
	16:0	16:1	16:3	18:0	18:1	18:2	18:3
Plasma	30–45	1–2		7–8	5–10	15–40	5–20
Mit inner	25–50			4–5	5–20	10–60	13–18
Mit outer	13–15			1–2	1–10	13–68	17–72
ER	20–30			5–6	2–6	28–50	4–41
Chloroplast	10–20	3–10	1–10	<1	1–5	3–8	65–75

[a] Abbreviations: Mit., mitochondrial; ER, endoplasmic reticulum.

Table 12. Species variation of phospholipid fatty acids

Fatty acid	% of total phospholipid fatty acids		
	Rat	**Sheep**	**Dog**
16:0	20–30	15–20	15–20
18:0	10–20	5–10	20–25
18:1	10–20	45–50	10–15
18:2	5–10	10–15	10–15
20:4	10–25	2–10	30

2.3. Glycolipids

Glycosylated derivatives of diacylglycerol (4, 5, 12–14)
These glycolipids (see *Figure 3*) are found in the membranes of bacteria and plants, but very rarely in animal membranes. In plants the glycosyl residues consist of usually one or two residues, but sometimes as many as four, linked to the free 3′-OH group. The most common are mono- and digalactosyl diacylglycerols, although sometimes glucose is present.

A sulfated derivative, sulfoquinovosyldiglyceride is also present in plants. The glycolipids of plants are heavily concentrated in the chloroplast membranes: *Table 13* gives a typical glycolipid profile. The only glycolipid of this type found in mammals is galactosyl dipalmitoylglycerol, in which the 3′ position of the galactose is sulfated: this is called seminolipid and is abundant in the testis and in semen.

Bacteria also have glycolipids of this type but the glycosyl groups can be more complex. They may also contain glycosyl derivatives of diacylglycerophosphate (phosphatidic acid) such as phosphatidyldiglucosyldiglyceride and glycerylphosphoryldiglucosyldiglyceride (see *Figure 4*). Another group of bacterial glycolipids is based on PI in which one or more of the

Table 13. Glycolipid profile of chloroplast membranes

Membrane	% of total glycolipid		
	MGDG[a]	**DGDG**[b]	**SQDG**[c]
Chloroplast envelope	20–35	55–70	8–15
Chloroplast lamellae	55–65	30–35	8–10

[a] MGDG = monogalactosyldiglyceride.
[b] DGDG = digalactosyldiglyceride.
[c] SQDG = sulfoquinovosyldiglyceride.

Figure 3. Glycolipids: glycosyl diglycerides. A, monogalactosyldiglyceride; B, digalactosyldiglyceride; C, sulfoquinovosyldiglyceride.

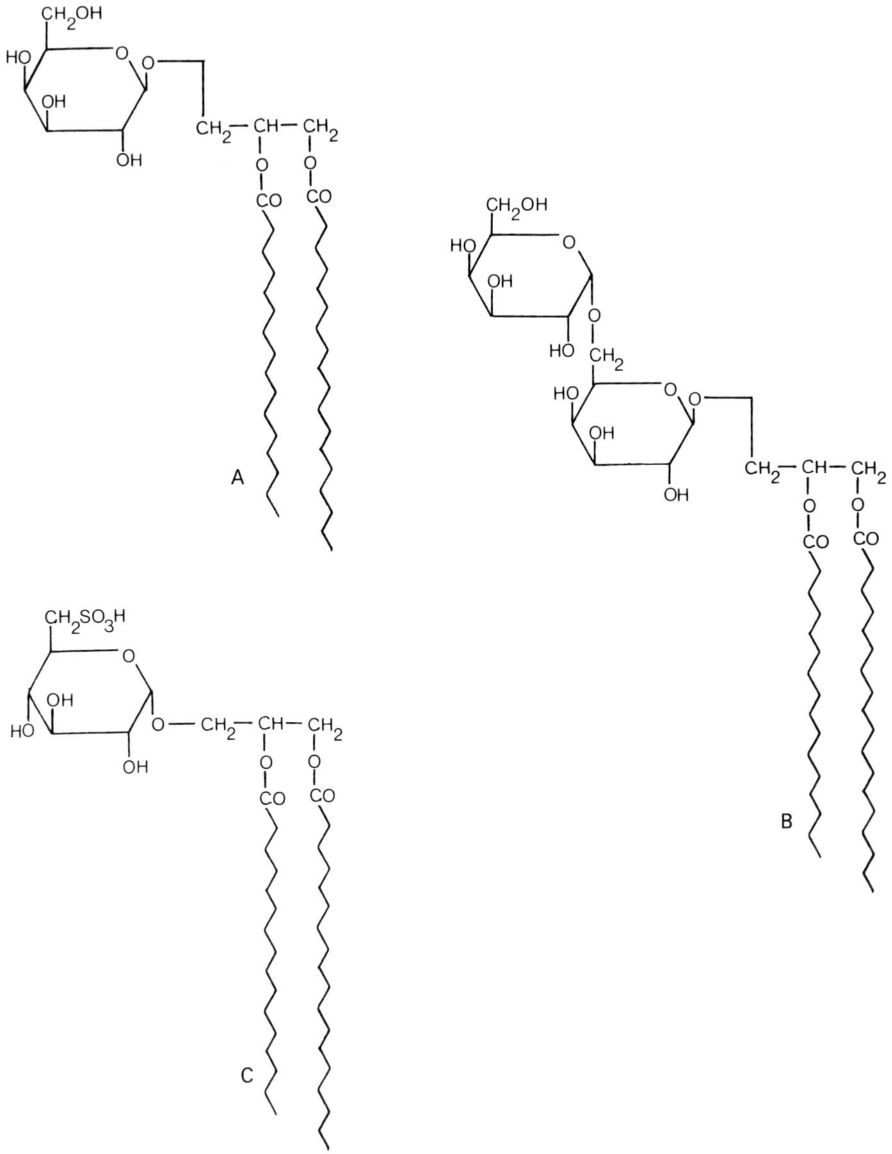

OH groups of the inositol ring is mannosylated. The glycolipid profile of the cytoplasmic membranes of bacteria is very specific to the organism and it is not useful to give a generalized table. *Mycobacter*, for example, contains 36% of PI-mono and -oligosaccharides, while other organisms contain only glycosyl diglycerides.

Glycosphingolipids (15–17)
Like all sphingosine-based lipids, these glycolipids are only found in animal membranes. They consist of ceramide linked (via the free —CH_2OH group) to either a straight- or

Figure 4. Bacterial glycolipids. A, glycerylphosphoryldiglucosyldiglyceride; B, phosphatidyldiglucosyldiglyceride.

A

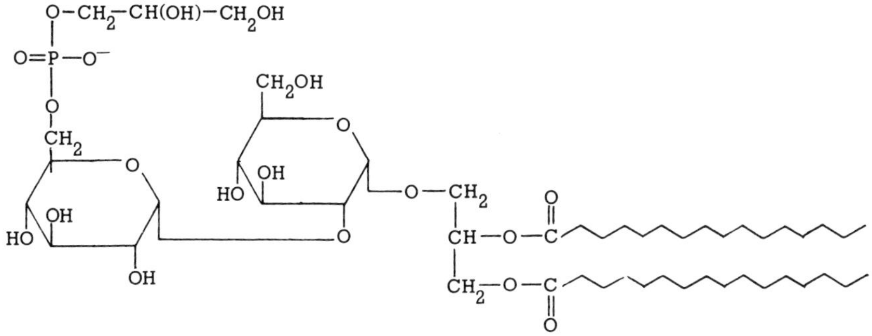

B

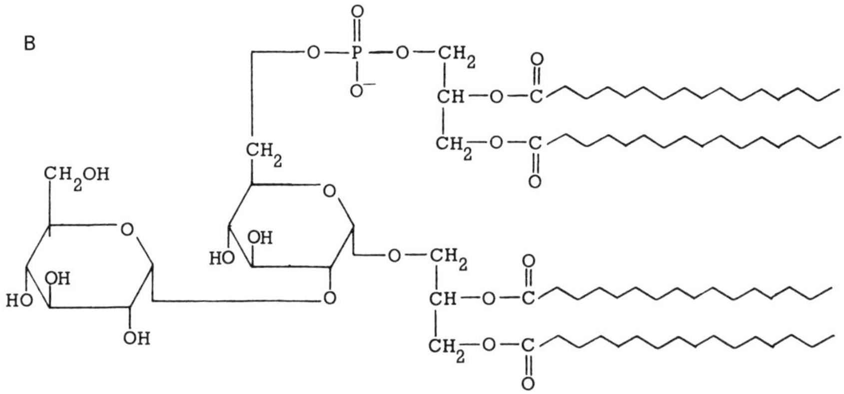

branched-chain oligosaccharide, which often includes *N*-acetylated sugars and *N*-acetylneuraminic acid (NANA) or sialic acid. Most commonly, the first sugar is glucose (although in a few cases this is galactose).

There are three broad groups based on the nature of their oligosaccharide chain:

(i) short (fewer than five neutral residues) unbranched chains (see *Figure 5*);

(ii) long (five or more neutral residues) mainly branched chains. This group includes blood group active glycosphingolipids and some with 20–50 residues have been isolated;

(iii) gangliosides (frequently branched and containing NANA), normally the number of saccharide residues is 3–7 (see *Figure 5*).

Information on the relative distribution of glycolipids in animal cell membranes is largely confined to a description of the molecular types in whole cells (where they are heavily concentrated in the plasma membrane) and how this profile changes in various disease situations. In myelin the most common glycosyl ceramide is galactosyl ceramide, which is sometimes sulfated on the C3 position of the sugar. In other plasma membranes, glucosyl ceramide predominates. Many of the gangliosides are also enriched in myelin.

Figure 5. Glycosphingolipids. A few examples of the range of glycosphingolipids are given: Gal, galactose; Glc, glucose; Cer, ceramide; GlcNAc, *N*-acetylglucosamine; GalNAc, *N*-acetylgalactosamine; NANA, *N*-acetylneuraminic acid.

Structure	Name
Galβ1-1Cer	Galactosylceramide
Gal1-4Galα1-4Galβ1-1Cer	Galactriaosylceramide
Galβ1-4Glcβ1-1Cer	Lactosylceramide
Galα1-4Galβ1-4Glcβ1-1Cer	Globotriaosylceramide
GalNAcβ1-3Galα1-3Galβ1-4Glcβ1-1Cer	Isoglobotetraosylceramide
Galβ1-4GlcNAcβ1-4Galβ1-4Glcβ1-1Cer	Gangliotetraosylceramide
Galβ1-3GalNAcβ1-4Galβ1-4Glcβ1-1Cer 3 | NANAα2	Ganglioside G_{M1}
3GalNAcβ1-4Galβ1-4Glcβ1-1Cer 3 | NANAα2	Ganglioside G_{M2}
NANAα2-3Galβ1-4Glcβ1-1Cer	Ganglioside G_{M3}
NANAα2-3Galβ1-3GalNAcβ1-4Galβ1-4Glcβ1-1Cer 3 | NANAα2	Ganglioside G_{D1a}
NANAα2-3Galβ1-3GalNAcβ1-4Galβ1-4Glcβ1-1Cer 3 | NANAα2-8NANAα2	Ganglioside G_{T1b}

2.4. Sterols

The only other major lipid in eucaryotic membranes is sterol, which is generally absent from procaryotes. In animals the major sterol is cholesterol (18), in plants either stigmasterol or sitosterol (see *Figure 6*). It is present almost entirely as the free sterol. In animal cells the nuclear membrane may contain some cholesterol ester but its precise location and its significance is not clear. Unlike the free sterol, it is not amphipathic so it will not orientate in the bilayer. Any ester in Golgi membranes is likely to be due to the presence of lipoproteins.

In both animal and plant cells the sterol is concentrated very much in the surface membrane. The other organelles where it is present to more than 10% of the total lipids are the lysosomes and perhaps subfractions of the Golgi, locations which are probably involved in the cycling of surface membrane components.

2.5. Lipid asymmetry

Evidence, mainly from work with the human erythrocyte membrane, shows that individual phospholipids and glycolipids are not uniformly distributed between the two halves of the bilayer (see *Table 14*). For more information on lipid asymmetry see ref. 3.

Figure 6. Animal and plant sterols. A, cholesterol; B, sitosterol; C, stigmasterol.

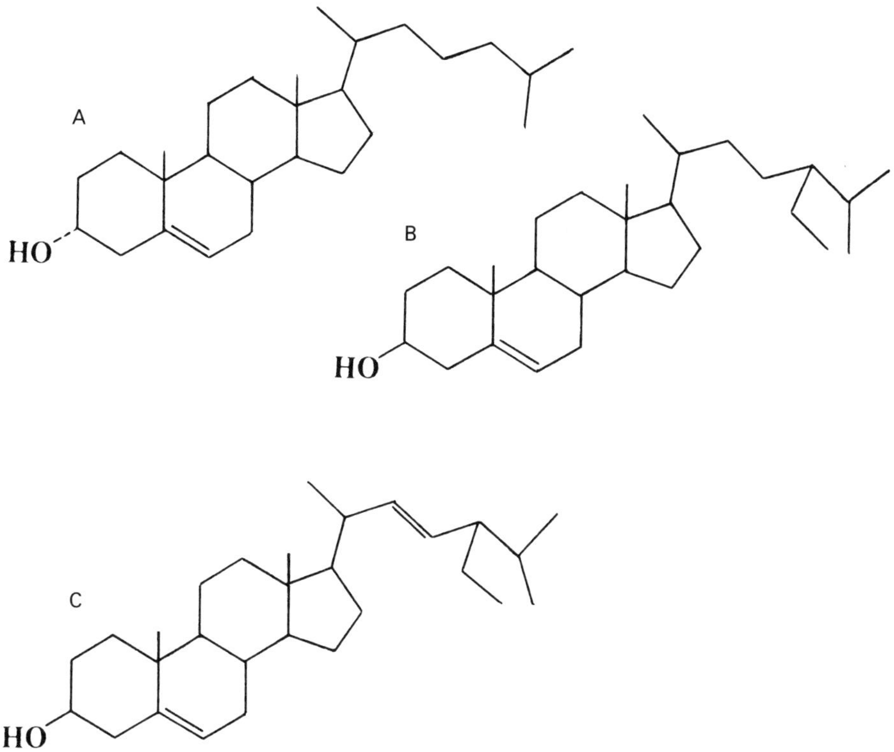

A

HO

B

HO

C

HO

3. MEMBRANE PROTEINS

3.1. General disposition

Most of the detailed knowledge regarding the general disposition of proteins in membranes has come from work on eucaryotic plasma membranes and from procaryotic cytoplasmic membranes, and to a much lesser extent on other membranes.

Table 14. Lipid asymmetry in the plasma membrane

Lipid type	Inside	Outside
Sphingomyelin	17	83
Phosphatidylcholine	26	74
Phosphatidylethanolamine	77	23
Phosphatidylserine	95	<5
Glycolipids	<5	95

Membranes

Figure 7. Disposition of proteins in membranes (see text for details).

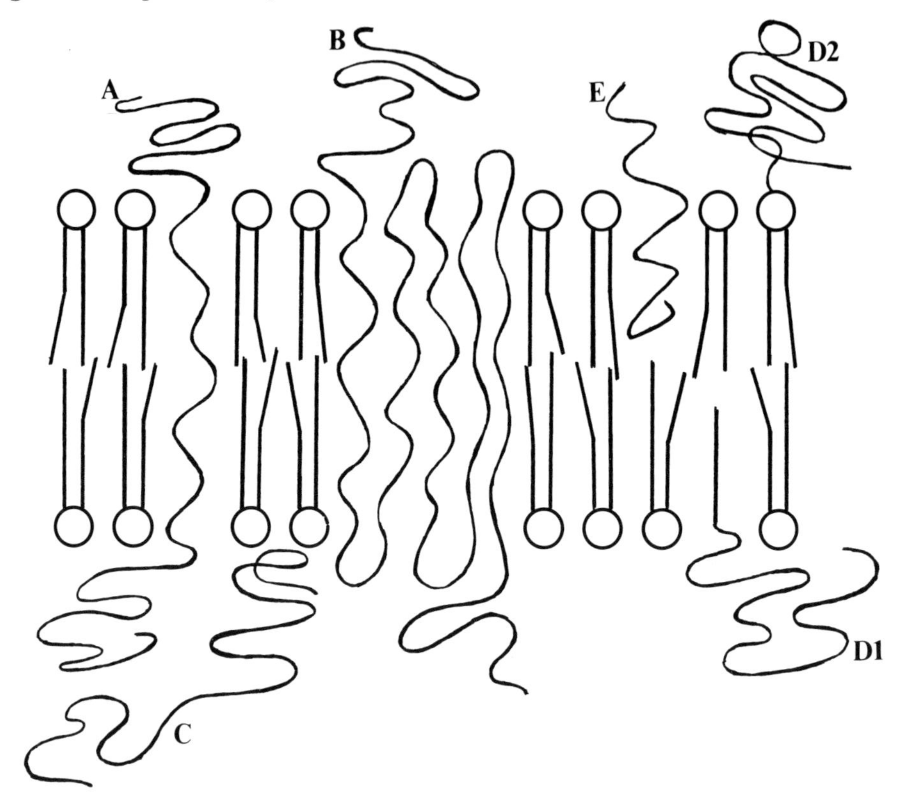

There are at least five categories of proteins in the eucaryotic plasma membrane with distinctive spatial arrangements and/or links to the membrane lipid bilayer (see *Figure 7*).

Type A proteins
The polypeptide chain passes once completely across the lipid bilayer. These integral (transmembrane) proteins have three distinct domains: two predominantly hydrophilic domains, extending outwards from the polar head groups of the amphipathic lipids on both the cytoplasmic and extracellular faces of the membrane, separated by a predominantly hydrophobic domain within the central lipid bilayer hydrocarbon region. This transmembrane segment is α-helical. The extracellular hydrophilic domain is usually glycosylated on one or more residues (asparagine, serine or threonine). The C-terminus is usually, but not invariably, on the cytoplasmic side.

Type B proteins
The polypeptide chain passes several times (up to 12, but seven is common) across the membrane bilayer. These proteins have multiple domains:
(i) several (x) predominantly hydrophobic α-helical domains which span the central hydrocarbon region of the lipid bilayer;
(ii) ($x-1$) predominantly hydrophilic domains which link consecutive transmembrane domains; and
(iii) two terminal hydrophilic domains which, depending on the number of transmembrane segments, may either be in the same or in different aqueous compartments (extracellular or cytoplasmic).

The hydrophilic linking segments may be short (as few as six residues) hairpins, longer hairpins (e.g. 20 residues) or rather more extensive loops (e.g. 60 residues). Glycosylation of any of the hydrophilic domains on the extracellular side is common.

Type C proteins

These are extrinsic proteins. They are essentially similar to soluble globular proteins which are bound ionically to the charged head groups of amphipathic lipids or to the hydrophilic domains of Type A or B proteins.

Type D proteins

These are proteins which are anchored to the membrane bilayer by a hydrophobic lipid tail.

Type D1 proteins

The lipid link is a fatty acid residue linked through its carboxyl group to an amino acid side-chain. A common fatty acid is myristate. These fatty acyl anchors can also occur on trans-membrane segments of Type A proteins.

Type D2 proteins

A more complex lipid anchor involves the covalent attachment of the C-terminus of the protein, via a phosphorylethanolamine residue, to a glycolipid which is a heavily glycosylated derivative of PI. This glycolipid is similar to one that is found in certain bacteria (see Section 2.3). As with the bacterial glycolipid, mannose residues are common in the 'glycan' link between the protein-ethanolamine phosphate and the PI, but in the eucaryote other residues such as galactose and glucosamine are also present (see *Figure 8*).

Type E proteins

Proteins that are partially buried within one half of the bilayer are often depicted in membrane models, but evidence for their existence is very limited. If they exist at all, it might be expected that their amino acid sequences would exhibit hydrophobic domains (like Type A and B proteins) in regions adjacent to the hydrocarbon chains of the bilayer.

G-proteins were thought to fall into this category, but their amino acid sequences have failed to reveal such domains and these important molecules are now thought to be Type D1 proteins (19). The only peptide that does, apparently, insert into one half of the lipid bilayer is the bee-venom active agent, mellitin; but since this acts as a membrane perturbant it cannot be considered a typical membrane protein.

Figure 8. Structure of a protein with a lipid anchor (Type D2). Man, mannose; Gal, galactose; $GlcNH_2$, glucosamine.

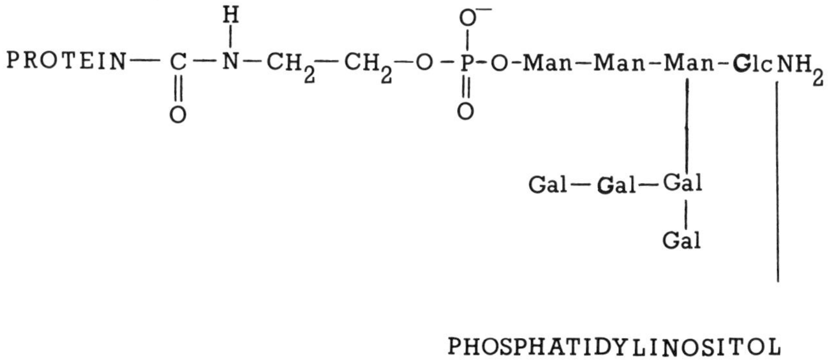

```
                 H              O⁻
                 |              |
PROTEIN— C —N—CH2—CH2—O—P-O-Man—Man—Man-GlcNH2
         ||                     ||              |        |
         O                      O               |        |
                                                |        |
                                   Gal— Gal— Gal         |
                                                |        |
                                               Gal       |
                                                         |

                                PHOSPHATIDYLINOSITOL
```

3.2. General functions

Type A proteins

(i) Receptors that are involved in the internalization of the ligand.
(ii) Receptors linked to cytoplasmic signaling.
(iii) Some antigen receptors.
(iv) Communication links between extracellular and intracellular proteins, particularly between proteins of the cell coat and the cytoskeleton.

Type B proteins

(i) Transporters of ions or small water-soluble molecules (e.g. sugars).
(ii) Receptors that are ion channels.
(iii) Receptors linked to cytoplasmic signaling.
(iv) Energy transducers.

Often a protein of Type A or B will fall into more than one functional category, e.g. energy transducers that are coupled to ion transport.

Type C proteins

(i) Provide links between type A or B proteins and proteins of the cytoskeleton or extracellular matrix.
(ii) By interacting with a Type A or B protein to provide an essential functional component.

Type D1 proteins

Certain G-protein subunits.

Type D2 proteins

(i) Some specific enzyme proteins.
(ii) Antigen receptors.

Table 15 lists a number of proteins in terms of their disposition in the membrane. For more information of receptors and their function see ref. 3.

3.3. Domain sizes and orientation of protein Types A and B

The transmembrane domain of an integral protein is approximately 23 amino acid residues long; the size of the terminal extracellular and cytoplasmic domains is variable. The cytoplasmic domain(s) may be particularly extensive in receptors that have tyrosine kinase activity (20–25).

In Type B proteins that act as ion transporters (e.g. Na^{+}–K^{+} ATPase and Ca^{2+} ATPase) the links between transmembrane segments on the cytoplasmic side may be particularly extensive to accommodate the ATP binding and phosphorylation domains (about 500 residues in total in the Ca^{2+} ATPase).

The most common location for the N-terminus is extracellular, and cytoplasmic for the C-terminus; but the reverse is not uncommon, while in a few Type B proteins (those with an even number of transmembrane domains) both N- and C-termini may be cytoplasmic. The size and disposition of the various domains of some selected membrane proteins are given in *Table 16.*

The Na^{+}–K^{+} ATPase and Ca^{2+} ATPase ion transporters both have eight transmembrane domains, with both N- and C-termini on the cytoplasmic side.

Table 15. Disposition and function of some membrane proteins

Protein type	Protein	Activity
A	Epidermal growth factor (EGF) receptor	Activates tyrosine kinase
	Insulin receptor	Activates tyrosine kinase
	Transferrin receptor	Internalizes ferritin
	Low-density lipoprotein (LDL) receptor	Internalizes LDL
	Glycophorin (human erythrocyte)	Links to spectrin
	Integrins	Link extracellular (e.g. fibronectin) and cytoskeletal proteins (e.g. talin)
B	Nicotinic acetylcholine receptor	Modulates Na^+ channels
	γ-Aminobutyric acid (GABA) receptor	Modulates Cl^- channels
	β-Adrenergic receptor	G-protein/adenylate cyclase
	Na^+–K^+ ATPase	Exchanges Na^+ and K^+ across the plasma membrane
	H^+ ATPase	Pumps protons across membranes
	Ca^{2+} ATPase	Pumps Ca^{2+} into ER[a] cisternae
	Rhodopsin	G-protein/Na^+ channel modulation
	Bacteriorhodopsin	Proton pump
	ATP synthetase (F_0 subunit)	Energy transducer/proton transport
	PSII complex (chloroplasts)	Energy transducer/proton transport
	lac Permease (*E. coli*)	Proton/β-galactoside transport
	Glucose transport protein (mammals)	Glucose transport into cell
C	Cytochrome *c*	Transfers electrons from complex II to complex III
	F_1 subunit of ATP synthetase	Phosphorylation site
	α-Subunit of insulin receptor	Ligand binding
	Protein 4.1 (erythrocyte membrane)	Links glycophorin to spectrin
D1	G-proteins	GTP/GDP binding
D2	5′-Nucleotidase	Hydrolyses AMP
	Alkaline phosphatase	Hydrolyses X-P
	Acetylcholinesterase	Hydrolyses acetylcholine
	Thy-1 receptor	Cell activation

[a] ER = endoplasmic reticulum.

Table 16. Size of domains of some integral membrane proteins

Protein	Number of transmembrane segments	Extracellular domain	Cytoplasmic domain
β-Adrenergic receptor	7	35(N)	85(C)
EGF receptor	1	620(N)	500(C)
Glucose transporter	12	15	40(N+C)
LDL receptor	1	750(N)	50(C)
Insulin receptor (β-subunit)	1	185(N)	400(C)
Transferrin receptor	1	675(C)	60(N)
Glycophorin	1	70(N)	40(C)
Bacteriorhodopsin	7	7(N)	25(C)
Bovine rhodopsin	7	40(N)	40(C)

3.4. Amino acid composition

Transmembrane domains (20–25)
The amino acids that predominate in the transmembrane domain are more hydrophobic than those in the remainder of the polypeptide chain. The domain is rich in non-polar amino acids, particularly leucine, isoleucine, valine, phenylalanine and tyrosine, while tryptophan, which is the most hydrophobic residue, is rather less common. Other non-ionized amino acids, such as methionine, alanine and glycine, can also occur (see *Table 17*).

Transmembrane domains of Type B proteins, particularly of transport proteins, may also contain residues with OH and $CONH_2$ groups, which may provide a polar channel within the array of transmembrane domains or binding sites (e.g. for glucose). The presence of proline in these domains of transport proteins may be associated with the ability of the protein to undergo conformational changes and so cause the opening and closing of channels (20).

Cytoplasmic domains
Although the amino acid composition of these domains is very variable overall, there are some sequences which seem to be common to many Type A and B proteins (20–25).

The residues in the cytoplasmic domain immediately adjacent to the transmembrane domain are often highly basic in Type A proteins (see *Table 18*).

These positively charged sequences are less well defined in Type B proteins, but even in these proteins, basic amino acid residues are common to one or more of the cytoplasmic loops. In the β-adrenergic receptor, six out of a 15-residue sequence of the third loop are basic.

Table 17. Amino acid composition of transmembrane domains

Amino acid	Glycophorin	Insulin	IgM	β-Adrenergic receptor						
				I	II	III	IV	V	VI	VII
Tryptophan	—	—	1	—	1	1	2	—	1	1
Phenylalanine	1	4	4	1	1	1	1	2	3	1
Tyrosine	—	1	1	—	—	—	1	2	—	3
Leucine	3	4	6	2	4	3	3	1	3	3
Isoleucine	4/5	6	1	5	2	2	4	3	5	3
Valine	2/3	3	2	5	3	4	3	5	3	2
Methionine	1	—	—	2	1	—	2	1	1	—
Alanine	1/2	—	—	2	5	2	—	3	—	—
Glycine	3/4	2	—	2	1	—	1	—	—	—

Serine and threonine residues are common close to the C-terminus, particularly if the protein is involved in phosphorylation (e.g. rhodopsin and the β-adrenergic receptor). This is not the case with the insulin receptor whose phosphorylation site is a tyrosine.

Extracellular domain (20–25)
Cysteine-rich regions are common, some of which are highly localized and others are more dispersed. Asparagine, serine and threonine residues may also be concentrated in this domain as sites for glycosylation.

There are 15 glycosylated serine and threonine residues and an asparagine residue in glycophorin. The EGF receptor has 12 glycosylation sequences (Asn–X–Ser/Thr). The α-subunit of insulin has 13 glycosylated residues and the β-subunit four such residues.

Table 18. Basic amino acid sequence of cytoplasmic domain close to lipid bilayer

Protein	Basic sequence
Insulin receptor	Arg—Lys—Arg
EGF receptor	Arg—Arg—Arg
LDL receptor	Lys—Asn—Trp—Arg—Leu—Lys
Glycophorin	Arg—Arg—Leu—Ile—Lys—Lys

3.5. Oligosaccharides of glycoproteins

Locations of glycoproteins in the cell

Glycoproteins are found predominantly in the plasma membrane, where the oligosaccharide chains are exclusively in the extracellular space. The polypeptide chains of receptors (Type A and B proteins) are invariably glycosylated, as are many extrinsic proteins (Type C, e.g. the α-subunit of insulin). *N*-glycosylated sequences are linked to asparagine (Asn), *O*-linked sequences to serine (Ser) or threonine (Thr).

Core sugars

The oligosaccharide units consist of core sugars and variable peripheral sequences (26). The core structure of *N*-linked oligosaccharides is given in *Figure 9*. The branched core oligosaccharide (2GlucNAc and 3Mann) of *N*-linked structures is derived from a parent core which is both more highly mannosylated and more branched (see *Figure 9*).

The fate of this parent core is determined by the polypeptide to which it is attached. It may be trimmed enzymically to yield the core for a plasma membrane glycoprotein, or it may be phosphorylated to yield a glycoprotein destined for lysosomes (*Figure 9*).

The parent core itself is formed in the Golgi from a more highly glycosylated precursor core which is transferred to the asparagine residue from dolichol phosphate in the smooth ER (*Figure 9*).

There are five major core sugar sequences of *O*-linked oligosaccharides in which the sugar linked to Ser or Thr (GalNAc) is common to all:

(i) a monosaccharide, GalNAc;
(ii) a disaccharide, GalNAc–Gal;
(iii) a disaccharide, GalNAc–GlcNAc;
(iv) a branched trisaccharide, GalNAc–(Gal, GlcNAc);
(v) a branched trisaccharide, GalNAc–(GlcNAc, GlcNAc).

Like the *N*-linked oligosaccharides, the elaboration of the core sugar structure occurs sequentially in the smooth ER and the Golgi membranes. The stepwise addition of all peripheral sugar residues to the core takes place in the Golgi (probably in the median to *trans* region).

Structure of glycoprotein oligosaccharides

N-linked oligosaccharides. Figure 9 shows some of the peripheral sugar sequences that may occur. The main residues are Gal, GlcNAc, NANA and fucose. Some contain repeat sequences of a Gal, GlcNAc disaccharide. More extensive and less commonly, the oligosaccharide chain of N–CAM contains 20–200 NANA residues.

O-linked oligosaccharides. Figure 10 shows some of the peripheral sugar sequences that may occur. The most common core sequence is probably the second. Each core type is capable of accepting a huge variety of peripheral sugar sequences; the sugar sequences appear to be controlled by the polypeptide.

3.6. Proteoglycans

Proteoglycans consist of long, unbranched carbohydrate chains, usually linked via a serine residue to a short polypeptide chain, which may include a hydrophobic sequence for anchoring to the membrane. The carbohydrate chains are frequently 40–100 residues in length and they are called the glycosaminoglycan chains. They reside predominantly in the

► p. 100

Figure 9. *N*-linked oligosaccharide sequences. Asn, asparagine; GlcNAc, *N*-acetylglucosamine; Gal, galactose; Fuc, fucose; NANA, *N*-acetylneuraminic acid (sialic acid); Glc, glucose. Structure 1, core oligosaccharide is used as an acceptor for plasma membrane glycoprotein peripheral sequences. Structure 2, plasma membrane glycoprotein oligosaccharide structure (short peripheral chains). Structure 3, plasma membrane glycoprotein oligosaccharide structure (long peripheral chains containing repeat sequences). Structure 4, parent oligosaccharide core; phosphorylated derivative directed to lysosomes. Structure 5, precursor oligosaccharide core; synthesized in the smooth ER; enzymic cleavage in the smooth ER and Golgi produces structures 4 and 1.

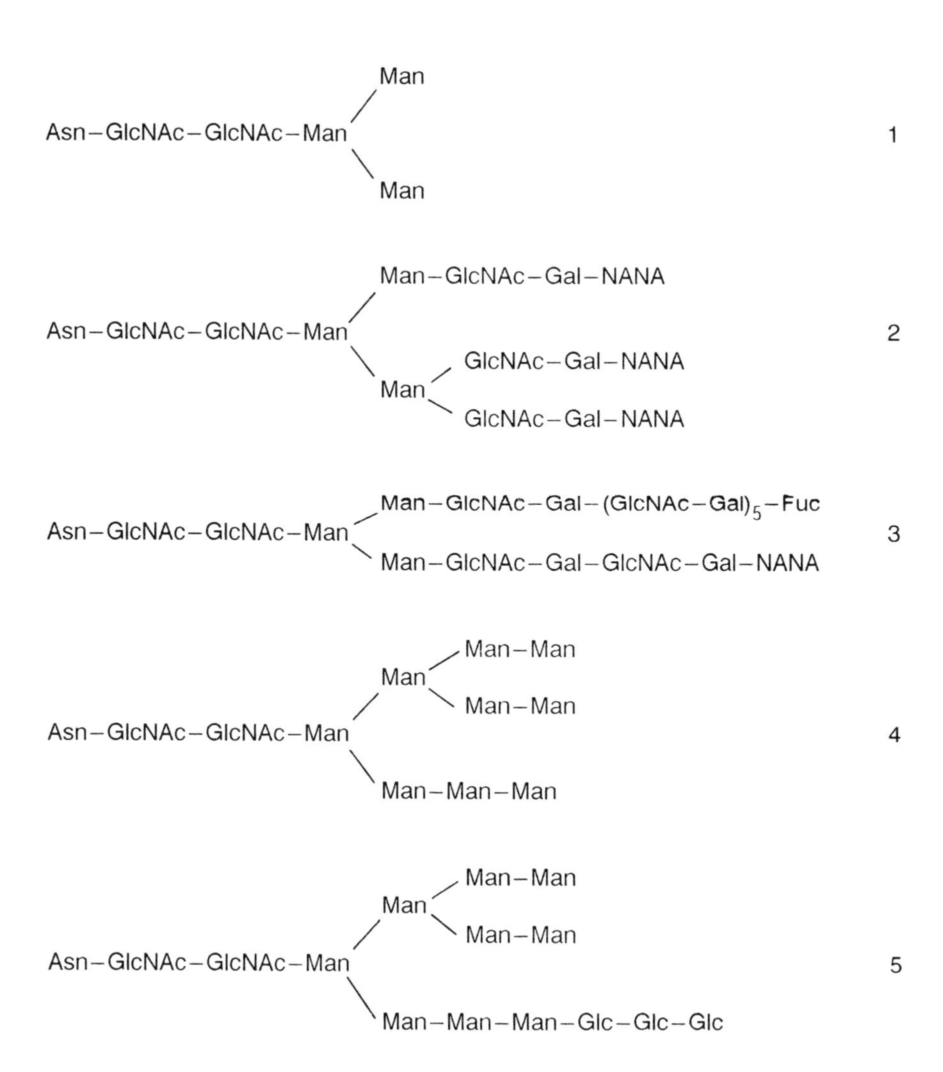

Figure 10. *O*-linked oligosaccharide sequences. Examples of peripheral sequences that may be linked to one of five core structures. Ser, serine; Thr, threonine; Gal, galactose; GlcNAc, *N*-acetylglucosamine; GalNAc, *N*-acetylgalactosamine; Fuc, fucose; NANA, *N*-acetylneuraminic acid (sialic acid). The core oligosaccharide is boxed.

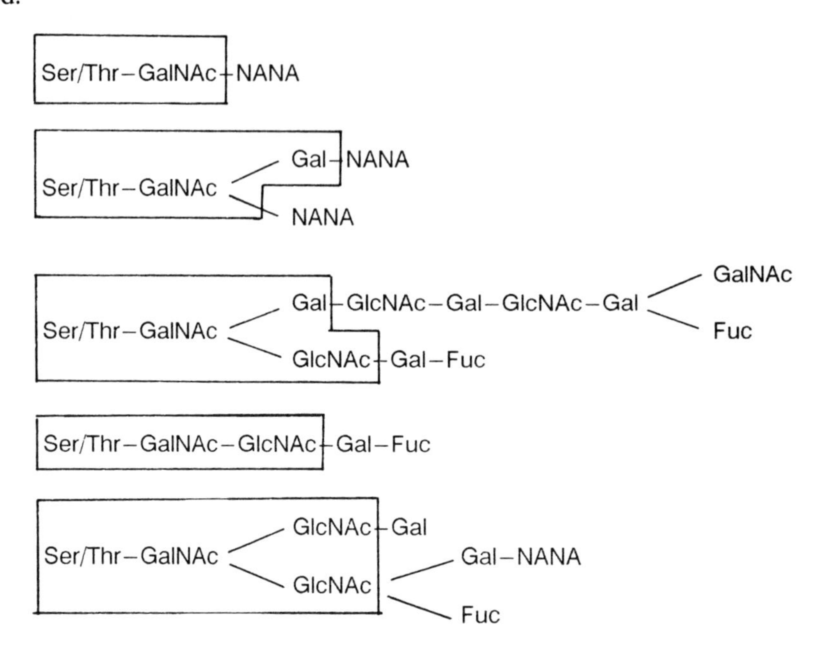

cell coat of cells (glycocalyx) where they are responsible for cell–cell adhesion and are involved in controlling growth and permeability.

The glycosaminoglycan chains commonly contain a repeating disaccharide unit, the precise molecular configuration of which may vary from chain to chain. *N*- or *O*-sulfated forms of uronic acids and amino sugars, occur frequently. Heparan sulfate, dermatan sulfate and chondroitin sulfate occur quite widely, along with more recently identified types such as fibronectin, vitronectin and laminin. These molecules vary in molecular size: they are commonly between 150 and 450 kDa, but may be as large as 900 kDa (27–29).

4. REFERENCES

1. Jain, M.K. (1988) *Introduction to Biological Membranes (2nd edn)*. John Wiley and Sons, New York.

2. Datta, D.B. (1987) *A Comprehensive Introduction to Membrane Biochemistry*. Floral Publishing, Madison.

3. Evans, W.H. and Graham, J.M. (1989) *Membrane Structure and Function*. IRL Press at Oxford University Press, Oxford.

4. Harwood, J.L. and Walton, T.J. (1988) *Plant Membranes — Structure, Assembly and Function*. Biochemical Society, London.

5. Rogers, H.J., Perkins, H.R. and Ward, J.B. (1980) *Microbial Cell Walls and Membranes*. Chapman and Hall, London.

6. Hawthorne, J.N. and Ansell, G.B. (1982) *Phospholipids*. Elsevier North Holland, Amsterdam.

7. Colbeau, A., Nachbaur, J. and Vignais, P.M. (1971) *Biochim. Biophys. Acta*, **249**, 462.

8. Mangold, H.K. and Paltauf, F. (1983) *Ether Lipids — Biochemical and Biomedical Aspects*. Academic Press, New York.

9. Berridge, M.J. (1987) *Ann. Rev. Biochem.*, **56**, 159.

10. De Pont, J.J.H.H.M., van Prooijen-van Eeden, A. and Bonting, S.L. (1978) *Biochim. Biophys. Acta*, **508**, 464.

11. Mandersloot, J.G., Roelofsen, B. and de Gier, J. (1978) *Biochim. Biophys. Acta*, **508**, 478.

12. Harwood, J.L. (1980) in *The Biochemistry of Plants* (P. K. Stumpf ed.). Academic Press, New York, Vol. 4, p. 1.

13. Goren, B.M. (1972) *Bacteriol. Rev.*, **36**, 33.

14. Shaw, N. (1970) *Bacteriol. Rev.*, **34**, 365.

15. Curatolo, W. (1987) *Biochim. Biophys. Acta*, **906**, 111.

16. Curatolo, W. (1987) *Biochim. Biophys. Acta*, **906**, 137.

17. Kannagi, R., Watanabe, K. and Hakomori, S. (1987) *Methods Enzymol.*, **138**, 3.

18. Yeagle, P.L. (1985) *Biochim. Biophys. Acta*, **822**, 267.

19. Simon, M.I., Strathmann, M.P. and Gautain, N. (1991) *Science*, **252**, 802.

20. Dohlman, H.G., Caron, M. G. and Lefkowitz, R.J. (1987) *Biochemistry*, **26**, 2657.

21. Goldfine, I.D. (1987) *Endocrin. Revs*, **8**, 472.

22. Ullrich, A., Bell, J.R., Chen, E.Y., Herrara, L.M., Petruzelli, L.M., Dull, T.J., Gray, A., Coussens, L., Liao, Y.-C., Tsubokawa, M., Mason, A., Seeburg, P.H., Grunfeld, C., Rosen, O.M. and Ramachandran, J. (1985) *Nature*, **313**, 756.

23. Neer, E.G. and Clapham, D.E. (1988) *Nature*, **333**, 126.

24. Stratford, M.M. and Cuatrecasas, P. (1985) *J. Membr. Biol.*, **88**, 205.

25. Strange, P.G. (1988) *Biochem. J.*, **249**, 309.

26. Paulson, J.C. (1989) *Trends in Biochem. Sci.*, **14**, 272.

27. Gallagher, J.T., Lyon, M. and Steward, W.P. (1986) *Biochem. J.*, **236**, 313.

28. Poole, A.R. (1986) *Biochem. J.*, **236**, 1.

29. Gallagher, J.T. (1989) *Curr. Opin. Cell Biol.*, **1**, 1201.

CHAPTER 7
PROPERTIES OF MEMBRANE TRANSPORT MECHANISMS

P. D. Brown and A. C. Elliott

1. TRANSPORT OF NON-ELECTROLYTES ACROSS BIOLOGICAL MEMBRANES

The major constituents of all biological membranes are lipids (see Chapter 6) and, consequently, lipid-soluble substances can dissolve in the membrane and diffuse across quite easily. The permeability of membranes to lipid-soluble compounds is directly proportional to their partition coefficient in oil (a single compound has a range of permeabilities in different membranes due to variations in the lipid composition of those membranes). The direction of movement is entirely dependent on the concentration difference across the membrane, and the rate of transport increases in a linear manner with increases in the concentration difference. Diffusion through the membrane is therefore described by Fick's Law (see *Equation 1* and the Appendix, *Equation 3*):

$$J = \frac{D_m(C_0 - C_i)}{x} \qquad \text{Equation 1}$$

where J is the rate of transport (moles cm^{-2} s^{-1}); D_m is the diffusion coefficient for the compound in the membrane (cm^2 s^{-1}); x is the membrane thickness (cm); and C_o and C_i are the concentrations of the molecule on the outside and inside of the cell (moles cm^{-3}).

Many physiologically important compounds (mostly small molecules) move across cell membranes by simple diffusion, for example carbon dioxide, oxygen and urea. However, other substances that are required by cells in quite large amounts, for example glucose, amino acids and nucleosides, are not particularly lipid soluble, mainly because they carry a slight electrical charge. The movement of these substances across biological membranes is therefore facilitated by specific proteins called membrane transport proteins, so that the rate of transport is much greater than that predicted on the basis of lipid solubility. Most non-electrolytes move by a process called facilitated diffusion. However, there are also more specialized Na^+-coupled co-transport processes for glucose and amino acids (see Section 2.3).

1.1. Facilitated diffusion

Facilitated diffusion is so called because the direction of transport is determined by the concentration gradient across the membrane (i.e. it is a passive process, as is diffusion). The rate of transport, however, increases in a non-linear fashion with an increase in the concenration gradient; that is, the carrier proteins are saturable and display Michaelis–Menten kinetics (see Appendix, *Equation 4*). There are specific transport proteins for different molecules (or groups of molecules). For example, Na^+-independent glucose transporters which transport a wide range of hexose and pentose sugars; Na^+-independent amino acid transporters (L for leucine and phenylalanine, y^+ for cationic amino acids, and T for tryptophan); and two types of nucleoside transporters for both adenosine and uridine. These transport

Membrane Transport

proteins also exhibit stereospecificity; for example, the glucose transporter in red blood cells transports D-glucose in preference to L-glucose.

2. TRANSPORT OF IONS ACROSS BIOLOGICAL MEMBRANES

By definition, ions are electrically charged and they are therefore relatively insoluble in the lipids of biological membranes. The movement of ions across membranes is therefore very dependent on ion-transport proteins. These proteins greatly enhance the rate of ion movement across membranes (e.g. the electrical resistance, a measure of ion permeability, of pure lipid bilayers is as high as 1×10^8 ohm cm^2, but that of a cell membrane may be as little as 5×10^3 ohm cm^2). They also make two other important contributions to the membrane: first, they bestow the property of ionic selectivity on the membrane (this is particularly important to cell homeostasis, cell signaling, etc.); and, secondly, some are able to transport an ion against its concentration gradient. There are three basic classes of ion-transport proteins: pumps, carriers or channels (see *Table 1*). To appreciate the fundamental differences between these classes of protein it is first necessary to consider the factors that influence ion movement across the cell membrane.

Table 1. Basic properties of transport proteins

	Molecular weight (kDa)	**Rate of transport (ion s^{-1})**	**Type of transport**
Channels	140–320	1×10^6–5×10^7	Passive
Carriers			
Co-transporters	75–200	5×10^0–4×10^3	Passive and secondary active
Exchangers	100	2×10^2–5×10^4	Passive and secondary active
Pumps	100–146	1×10^2–6×10^2	Primary active
Facilitated diffusion	55	6×10^2 [a]	Passive

[a] Units = molecules s^{-1}.

2.1. Factors influencing ion movements: electrochemical gradients

Ions in solution move by diffusion down chemical gradients, i.e. from regions of high concentration to regions of low concentration. However, as ions carry an electrical charge (cations are positive; anions are negative), they are also affected by electrical gradients; for example, the inside-negative potential difference found across all cell membranes favors the movement of cations into the cell. Thus, the movement of ions across cell membranes is influenced first by the chemical concentration gradient for the ion, and secondly by the electrical potential gradient across the membrane (see Appendix, *Equation 5*). The combined effect of these two driving forces, derived as the algebraic sum of the two forces, constitutes the electrochemical gradient for an ion. When the electrochemical gradient is zero (i.e. when the two forces on the ion are equal and opposite) the system is in a state of electrochemical equilibrium. The transmembrane potential at which this occurs, called the equilibrium potential (E in volts), is given by the Nernst equation (*Equation 2*):

$$E = \frac{RT}{zF} \cdot \ln \frac{C_o}{C_i} \qquad \text{Equation 2}$$

where R is the gas constant; T is temperature in K; z is the valency of the ion and F is Faraday's constant (see the Appendix for values). When the membrane potential is not equal to the equilibrium potential, the difference between the membrane potential and the equilibrium potential determines the net driving force on the ion (i.e. the direction and size of the electrochemical gradient).

The movement of ions via membrane transport proteins usually occurs down the electrochemical gradient for that ion (passive transport). There are certain classes of protein, however, that transport ions against the electrochemical gradient; this requires the expenditure of energy and is termed active transport. There are two distinct types of active transport: either primary active, in which the energy source molecule (ATP) is hydrolysed at the site of ion-transport (i.e. the membrane transport protein has ATPase activity), or secondary active, in which movement of an ion takes place on a protein which is not itself the site of ATP hydrolysis. Secondary active transport takes place because the transport of the ion is directly coupled to the movement of another species of ion (usually Na^+), which is transported (on the same protein) down its electrochemical gradient (a gradient that is created by primary ATP-driven active transport elsewhere in the membrane).

2.2. Ion channels

This is probably the most diverse group of ion-transport proteins in terms of function and possibly structure (although molecular structures are known for only a few ion channels). Transport of a single type of ion takes place in either direction through the channel, according to the electrochemical gradient for that ion. Channel activity can be modified acutely by extracellular or intracellular factors, so that more or fewer ions are transported depending on the needs of the cell. Most cells possess a number of channel types, which are permeable not only to different ion species, but also to ions of the same type; for example, cardiac muscle cells may contain as many as seven different types of K^+-selective channel.

Characteristics of ion channels

Passive. Ions move through channels down their electrochemical gradient.

Electrogenic. Net movement of electrical charge produces an electrical current which changes the transmembrane potential difference.

Selectivity. Always between cations and anions, frequently between ions with different charge (e.g. + versus 2+), and sometimes between ions of the same charge; for example, the channels in nerve axons discriminate very effectively between Na^+ and K^+ for the generation of action potentials.

Transport rate. This can be as high as 5×10^7 ions s^{-1}, which is several orders of magnitude higher than for other transport proteins. Transport activity is generally measured in electrical parameters; for example, conductance in pS, where 1 pS $\equiv 3 \times 10^4$ ions s^{-1} per 10 mV of applied potential difference (see Appendix).

Saturable kinetics. Conductance increases in an almost linear fashion with the concentration of the permeant ion in the physiological range, and only saturates at concentrations well in excess of physiological levels (i.e. 3×10^{-1} M for K^+ channels).

Regulation. Channel proteins fluctuate between 'open' and 'closed' states. Transport rate is regulated by a variety of physiological agents, which cause the channel to 'open' or 'close' more frequently (see *Table 2*).

Pharmacology. Channels can be blocked by impermeant ions, and by animal- or plant-derived toxins. The activity of some channels can also be increased by some drugs, for example Bay K 8644 on L-type Ca^{2+} channels or diazoxide on ATP-sensitive K^+ channels.

Importance. The diversity of channel types and regulating factors means that activation or inactivation of a particular ion channel can cause a specific change in cellular activity.

Classification of ion channels

The past decade has seen a rapid growth in our knowledge of channel diversity and regulation (mainly because of the development of the patch-clamp technique, which allows the properties of individual ion channels to be studied). There is not as yet, however, an established method for classifying channels. Various parameters can be used, for example conductance, selectivity, method of regulation, tissue/cell type of origin and pharmacology, and it is likely that any comprehensive classification scheme will need to make use of a combination of these. One of the most logical schemes, recently proposed by Eisenberg (1), classifies channels in terms of 'agonist–selectivity–conductance'; for example, a channel that is directly gated by acetylcholine, is selective for cations and has a conductance of 40 pS would be ACH–CAT–40.

2.3. Ion carriers

Ion carriers (*Table 3*) move two or more ions across membranes in each cycle of operation. There are two basic classes of ion-carrier proteins:

(i) Exchangers: ions of the same charge are transported in different directions across the membrane.
(ii) Co-transporters: ions of different charge (and some non-electrolytes) are transported in the same direction.

Carrier proteins perform many homeostatic functions for cells, for example volume regulation, maintenance of intracellular pH and accumulation of nutrients. There is little variation in these processes between cell types, so that there is generally little variety in carrier proteins; for example, three highly conserved genes may code for all the Cl^-–HCO_3^- exchange proteins found in many different cell types throughout the body.

Characteristics of ion carriers

Passive or secondary active. With passive ion carriers, all ions move down their (electro)chemical gradient. (The direction of electroneutral transport is not influenced by the electrical gradient, i.e. ions only move down their chemical gradient.) In the case of secondary active ion carriers, one (or more) ion moves down the (electro)chemical gradient and one (or more) ion moves against (up) the electrochemical gradient.

Table 2. Regulation of ion-channel activity[a]

Regulation	Description	Examples[b]
Voltage	Changes in the transmembrane potential cause the movement of groups of charged amino acids within the channel protein ('gates'), causing increased opening or closing of the channel	Na^+, K^+, Ca^{2+} channels in excitable tissues, e.g. nerve and muscle (opened by depolarization)
Ligand	Reversible binding to the ion-channel protein, resulting in increased opening or closing of the channel	
External	Neurotransmitters acting on some receptors at the neuromuscular junction and postsynaptic membranes	Acetylcholine: nicotinic receptor cation channel
		$GABA_A$ receptor Cl^- channel
		Glutamate receptor Na^+/Ca^{2+} channels
Internal	Intracellular second messengers released as a result of receptor activation (e.g. Ca^{2+} released from intracellular stores by muscarinic receptor activation)	Ca^{2+} activates K^+ and Cl^- channels in many secretory epithelia
		Inositol (1, 4, 5)tris-phosphate activates Ca^{2+}-release channels in intracellular Ca^{2+} storage organelles
Phosphorylation	Covalent modification of channel protein (addition of phosphate groups) through the action of protein kinases	cAMP-dependent protein kinase activates Cl^- channels in some secretory epithelia
		K^+ channels in nerves
		Ca^{2+} channels in cardiac myocytes

[a] This table is a summary of some well-known examples of channel regulation; it is not a comprehensive catalogue of all known ion channels.
[b] Abbreviations: GABA, γ-aminobutyric acid; cAMP, cyclic adenosine monophosphate.

Table 3. Ion carriers

	Direction of operation[a]	$K_{0.5}$ (M)[b]	Other ions carried	Inhibitors[c]	Occurrence and function
Exchangers					
Na^+–H^+	Na^+ – in H^+ – out	10^{-3}–10^{-2} 10^{-6}–10^{-5}	Li^+, NH_4^+	Amiloride and derivatives	Most cell types: intracellular pH (acid extrusion) and volume regulation
Cl^-–HCO_3^-	Cl^- – in HCO_3^- – out	3–9×10^{-3} 1–2×10^{-3}	Br^-, I^-, F^- SCN^-, NO_3^-, SO_4^{2-}	Stilbene derivatives (SITS, DIDS)	Many cell types: intracellular pH (acid accumulation) and volume regulation
Na^+-dependent Cl^-–HCO_3^-	Na^+ – in Cl^- – out HCO_3^- – in	10^{-1} *nd* 10^{-2}	—	SITS, DIDS	A few cell types: intracellular pH regulation (acid extrusion)
$3Na^+$–Ca^{2+}[d]	Na^+ – in Ca^{2+} – out	0.5–1×10^{-1} 0.2–1×10^{-5}	Sr^{2+}	Some amiloride derivatives (dichlorobenzamiloride) and Ni^+	Cardiac myocytes: Ca^{2+} extrusion
$4Na^+$–Ca^{2+}($-K^+$)[d]	Na^+ – in Ca^{2+} – out K^+ – in	0.5–1×10^{-1} 0.2–1×10^{-5} *nd*	Sr^{2+}	Some amiloride derivatives (dichlorobenzamiloride)	Excitable cells, e.g. rod cell in the retina: Ca^{2+} extrusion
Co-transports					
Na^+–$2Cl^-$–K^+	Na^+ – in Cl^- – in K^+ – in	5×10^{-3} 3–5×10^{-2} 10^{-2}	Li^+ $Br^- > NO_3^-$ $Tl^+ > Rb^+ > Cs^+ > NH_4^+$	Bumetanide, frusemide and other 'loop diuretics'	Epithelia, smooth muscle: Cl^- accumulation

Table 3. Continued

	Direction of operation[a]	$K_{0.5}$ (M)[b]	Other ions carried	Inhibitors[c]	Occurrence and function
Na^+–$nHCO_3^-$ [de]	Na^+ – in [out] HCO_3^- – in [out]	10^{-2} 10^{-6}–10^{-5}	—	DIDS, SITS	Neurones and glial cells: pH regulation (acid extrusion) [Kidney: HCO_3^- secretion]
Na^+-amino acid[d] (A, ASC, Gly systems)	Na^+ – in Amino acid – in	10^{-2} 10^{-4}–3×10^{-2}	—	—	Most cell types: nutrient uptake for cell metabolism
Na^+–glucose[df]	Na^+ – in Glucose – in	3–8×10^{-2} 1–5×10^{-4}	—	Phloridzin	Small intestine, kidney (proximal tubule): glucose absorption

[a] Direction of transport observed when ion concentrations are at or close to normal physiological concentrations. The direction of ion transport can be reversed if ion concentrations are experimentally manipulated.
[b] $K_{0.5}$ is the concentration of substrate at which the rate of transport is half the maximum transport rate. *nd*, Not yet determined.
[c] Abbreviations: SITS, 4-acetamido-4′-isothiocyanostilbene-2,2′-disulfonic acid; DIDS, 4,4′-diisothiocyanostilbene-2,2′-disulfonic acid.
[d] Electrogenic.
[e] There may be more than one type of Na^+–HCO_3^- transport. The direction of transport for these ions is dependent on the stoichiometry (number of ions carried) of the system, e.g. $1Na^+:3HCO_3^-$ is out of the cell, or $1Na^+:2HCO_3^-$ is into the cell.
[f] Na^+–glucose co-transport is just one example of a wide range of Na^+-coupled co-transporters for amino acids, carboxylic acids and bile acids found in the small intestine and kidney. These are responsible for the rapid and efficient absorption of nutrients from the lumen of the small intestine and the proximal tubule. NB: these transporters are not the same as the A and ASC systems that are found in most cells of the body.

Electroneutral. No net movement of charge; but there are exceptions, for example Na^+-glucose and Na^+-amino acid co-transport and Na^+–Ca^{2+} exchange which make use of the electrochemical gradient for Na^+.

Highly selective for charge and valency. Usually selective amongst ions of similar charge and valency.

Transport rate. Low, 5 to 5×10^4 ions s^{-1}.

Saturable kinetics for all transported ions. High affinity (e.g. $K_{0.5} = 10^6$ to 5×10^{-2} M).

Regulation. Operational as long as chemical gradients are present (unlike ion channels which exhibit inactive 'closed' states). Activity may be modified by:

(i) phosphorylation, for example Cl^-–HCO_3^- exchange in some types of epithelial cell;

(ii) ligand binding, for example Na^+–H^+ exchange has a H^+-binding site on the cytoplasmic side which, when occupied, enhances transport rates.

Pharmacology. Only a few, and generally not very specific, organic inhibitors have been identified for some of the carrier proteins (see *Table 3*).

Importance. Cell homeostasis (see above). In epithelia, in addition to the homeostatic roles, carriers have specific roles in net transepithelial ion transport (i.e. secretion and absorption). Cl^-–HCO_3^- exchange in red blood cells has an important role in the transport of CO_2.

2.4. Ion pumps

Ion pumps are the most specialized membrane transport proteins in that they use energy derived directly from ATP hydrolysis to transport ions against their electrochemical gradients. There are two classes of ion pump: E_1–E_2 ATPases in mammalian cell and organelle membranes; F_0–F_1 ATPases in mitochondrial, chloroplast and bacterial

▶ p. 112

Figure 1. Simplified kinetic scheme for Na^+–K^+ ATPase. E_1 and E_2 represent the two conformations of the pump protein. The single-headed arrows represent the conformational changes of the protein, and the double-headed arrows represent pump–ion association equilibria.

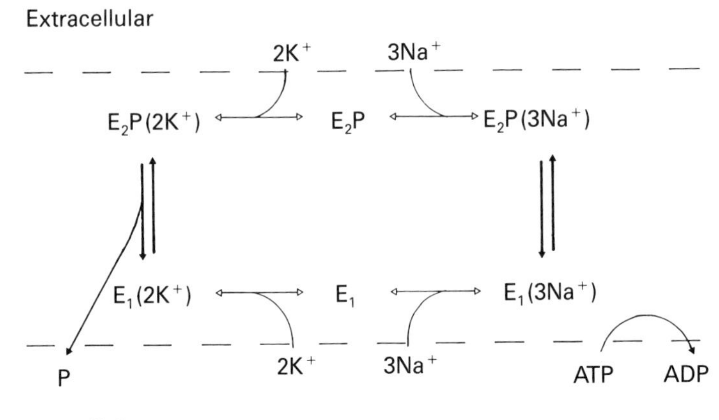

Table 4. Characteristics of ion pumps

Pump	Direction of operation[a] (ions per ATP)	$K_{0.5}$ (M)	Other ions transported	Inhibitors[b]	Occurrence and functions
Na^+–K^+ ATPase	3 Na^+ – out 3 K^+ – in	$1–5 \times 10^{-2}$ $1–2 \times 10^{-3}$	Li^+ Rb^+	Cardiac glycosides, e.g. ouabain, strophanthidin	All cells: K^+ accumulation Na^+ extrusion
Ca^{2+} ATPase (plasma membrane)	2 Ca^{2+} – out	$1–5 \times 10^{-7}$	–	Anti-calmodulins: pump activity is enhanced by calmodulins	All cells: Ca^{2+} extrusion
Ca^{2+} ATPase (organelles)	2 Ca^{2+} into organelle	$1–5 \times 10^{-7}$	—	Thapsigargin; 2,5 di(tert-butyl)-hydroquinone	Sarcoplasmic and endoplasmic reticulum: accumulation of Ca^{2+}
H^+–K^+ ATPase	2 H^+ – out 2 K^+ – in	$10^{-7}–10^{-6}$ $0.5–5 \times 10^{-3}$	— Tl^+, Rb^+, Cs^+	Omeprazole (a novel anti-ulcer drug)	Gastric parietal cells, colonic epithelial cells: H^+ secretion

[a] Hydrolysis of ATP causes ion movement in only one direction. If ion gradients are reserved experimentally the pump will generate ATP from ADP.
[b] All E_1–E_2 pumps are inhibited by vanadate ions.

Membrane Transport

membranes. This section considers only the E_1–E_2 ATPases found in mammalian cells; for example, Na^+–K^+ ATPase and Ca^{2+} ATPase (which are found in all cell types), and H^+–K^+ ATPase (which is found in a few specialized epithelial cells).

All the E_1–E_2 ATPases appear to have similar structural and functional characteristics. *Figure 1* is a diagrammatic representation of the mechanism of the Na^+–K^+ ATPase, which is representative of all E_1–E_2 ATPases. There are two physical conformations of the protein (E_1 and E_2). Ion transport across the membrane takes place at the same time as the conformational change. A distinctive feature of the E_1–E_2 ATPases is that the cycle of pump activity includes intermediary phosphorylated states, which do not occur in the cycle of activity of the F_0–F_1 ATPases.

Characteristics of E_1–E_2 ATPases

Primary active transport. Ions are transported against their electrochemical gradients at the expense of ATP hydrolysis.

Electrogenic. For example, Na^+–K^+ATPase, which transports three Na^+ ions for two K^+ ions. The exception is H^+–K^+ ATPase, which is electroneutral (see *Table 4*).

Highly selective. Even amongst ions of similar charge and valency.

Transport rate. Low, 10^2– 6×10^2 ions s^{-1}.

Saturable kinetics. $K_{0.5} = 10^{-6}$–10^{-2} M.

Regulation. The rate of transport is governed by the ion concentrations on either side of the membrane. The activity of cell membrane Ca^{2+} ATPase is also regulated by the cytoplasmic Ca^{2+}-binding protein, calmodulin. A rise in cytoplasmic Ca^{2+} activity causes an increase in Ca^{2+} binding to calmodulin, and thus an increase in Ca^{2+} ATPase activity.

Pharmacology. Vanadate ions bind to the phosphate-binding sites of all the E_1–E_2 ATPases, inhibiting further cycles of phosphorylation and dephosphorylation (vanadate does not inhibit F_0–F_1 ATPases). There are also specific organic inhibitors for some of the ion pumps (*Table 4*).

Importance. Na^+–K^+ ATPase activity accounts for between 30 and 60% of all ATP expenditure in many cells. Na^+–K^+ ATPase maintains:
(i) the Na^+ gradient across the cell membrane, which is needed to drive many other transport processes; and
(ii) the K^+ gradient, which is a major factor in determining the membrane potential.

Ca^{2+} ATPase and H^+–K^+ ATPase both produce 10^4- to 10^5-fold ion concentration gradients, which would be difficult to produce by secondary active transport.

3. REFERENCE

1. Eisenberg, R.S. (1990) *J. Memb. Biol.*, **115**, 1.

4. APPENDIX

4.1. Equations and constants for membrane transport

Diffusion through a membrane is described by *Equation 3*, which is a simplification of Fick's Law (see *Equation 1*).

$$J = P(C_o - C_i) \qquad \text{Equation 3}$$

where J is the rate of transport (moles cm^{-1} s^{-1});); P is the permeability coefficient for that membrane (cm s^{-1}); C_o and C_i are the concentrations outside and inside the cell (mole cm^{-3}).

Michaelis–Menten kinetics describe transport by some carrier system, as shown in *Equation 4*:

$$J = \frac{C \cdot J_{max}}{C + K_m} \qquad \text{Equation 4}$$

where J_{max} is the maximum rate of transport, C is the concentration of the substrate and K_m is the concentration of substrate at which J is exactly one-half of J_{max}.

Goldman–Hodgkin–Katz equation provides an estimate of the electrical potential gradient across the cell membrane (V_m):

$$V_m = \frac{RT}{F} \cdot \ln \frac{P_K[K]_o + P_{Na}[Na]_o + P_{Cl}[Cl]_i}{P_K[K]_i + P_{Na}[Na]_i + P_{Cl}[Cl]_o} \qquad \text{Equation 5}$$

where K^+, Na^+ and Cl^- are the permeable (contributing) ions: P_K, P_{Na} and P_{Cl} are the permeability coefficients for these ions; $[\ ]_o$ and $[\ ]_i$ are the extracellular and intracellular activities of these ions.

Constants

R (gas constant) $= 8.31$ J K^{-1} mol^{-1}

F (Faraday's constant) $= 96\,500$ C mol^{-1}

Avogadro's number $= 6.023 \times 10^{23}$ mol^{-1}

Charge on a single monovalent ion $= \frac{96\,500}{6.023 \times 10^{23}} = 1.6 \times 10^{-19}$ C

Membrane Transport

CHAPTER 8
CELL SURFACE RECEPTORS

D. Schulster

1. CHARACTERISTICS OF MEMBRANE CELL SURFACE RECEPTORS

Cell membrane receptors receive extracellular signals in the form of neurotransmitters and hormones (including growth factors) and communicate or transduce this information to the intracellular compartments (1). Although the characteristics of the membrane-bound receptors vary, they do have many properties in common. They are all glycoproteins composed of one or more polypeptide chains woven through the membrane lipid bilayer, and contain either one, four or seven transmembrane sequences that are α-helical and hydrophobic. The growth factor receptors have a single transmembrane domain (see *Table 1*, refs 2–34) and those receptors coupled to ion channels mostly have four transmembrane helices (see *Table 2*, refs 35–42). Those receptors transiently coupled to effector enzymes via GTP-binding proteins (G-proteins), have seven transmembrane helices (see *Table 3*, refs 43–58) and, of these, various subtypes have been reviewed recently (e.g. for the adrenoceptors (59), dopamine receptors (60) and histamine receptors (61)). For discussions of receptor subtype definitions and the arrangement of receptors into groups and superfamilies see ref. 60. Most cell surface receptors can be phosphorylated at domains protruding into the intracellular regions. They may be considered as allosteric proteins with functional states capable of interconversion, that may behave in a cooperative (or non-cooperative) manner (62–64).

The techniques of molecular biology have transformed the field of purification and characterization of receptors. Homologies and receptor groupings have now become apparent from the abundant structural data. Thus, as shown in *Tables 1–3*, it is now perceived that the growth factor and cytokine receptors comprise one main grouping, those receptors regulating ion channels comprise another and receptors coupled to G-proteins constitute a third group.

Within these three main groups several superfamilies have been perceived (see ref. 60 for discussion). Thus, the receptors given in *Table 1*, with a single, hydrophobic transmembrane domain, may be subdivided into the 'immunoglobin superfamily' (which includes receptors for interleukin-1, platelet-derived growth factor, T-cell antigens and other membrane glycoproteins; see ref. 65 for a review) and the 'hematopoietin superfamily' (which includes receptors for interleukins 2–4 and 6–7, granulocyte–macrophage colony stimulating factor, prolactin, growth hormone and erythropoietin; see ref. 66 for a review). The group of receptors coupled to G-proteins, given in *Table 3*, may be further subdivided into superfamilies of receptors coupled via the G-proteins to either phosphoinositide hydrolysis (*Table 4*), adenylate cyclase stimulation (*Table 5*) or adenylate inhibition (*Table 6*).

The general and cell-specific properties of the cell surface receptors have been comprehensively described (67) and this latter text provides the principles of receptor-binding assays and analysis of the binding curves obtained using radiochemical ligands (see also refs 68, 69), as well as details for receptor preparation and separation by classical biochemical methods.

Table 1. Classification of transmembrane receptors: examples of receptors with a single transmembrane segment per subunit (solubilized using Nonidet P-40 or Triton X-100)

Receptors for[a]	Structural and functional characteristics (: linked to)		Refs[b]
Insulin	Two α- (135 kDa) and two β- (95 kDa) subunits	(: tyrosine kinase)	2–6
IGF-I	Two α- (130 kDa) and two β- (90 kDa) subunits	(: tyrosine kinase)	6–9
IGF-II	Monomer (220–300 kDa)	(: G-protein)	10–12
NGF	Hetero-oligomer (83 kDa)	(: unknown)	13–15
EGF	Monomer (170 kDa)	(: tyrosine kinase)	6, 9, 16, 17
PDGF	Dimeric in active form? (180 kDa)	(: tyrosine kinase)	9, 10, 18–21
ANP	Monomer	(: guanyl cyclase)	22, 23
LDL	Monomer (precursor: 120 000 M_r) (mature protein: 95 000 M_r)	(: cholesterol uptake)	6, 24–26
IL-2	Hetero-oligomer (55 000–65 000 M_r)	(: tyrosine kinase)	27–29
FGF	Monomer (125 kDa and 145–160 kDa	(: tyrosine kinase)	9, 30
GH	Monomer (130 kDa and 50–60 kDa	(: unknown)	31–34

[a] Abbreviations: IGF-I, insulin-like growth factor-I; IGF-II, insulin-like growth factor-II; NGF, nerve growth factor; EGF, epidermal growth factor; PDGF, platelet-derived growth factor; ANP, atrial naturetic peptide; LDL, low-density lipoprotein; IL-2, interleukin-2; FGF, fibroblast growth factor; GH, growth hormone.

[b] References for receptor sequence and purification by affinity chromatography.

Table 2. Classification of transmembrane receptors: examples of receptors with four or more transmembrane segments per subunit functioning as receptor–ion channel complexes (solubilized using Triton X-100, sodium cholate or DOC)[a]

Receptors for[a]	Molecular weight	Structural and functional characteristics and purification methods	Refs[b]
n-ACh	290 kDa	Hetero-oligomers with at least 20 transmembrane segments per functional unit, acting as ion-channel complexes	35–37
$GABA_A$	220 kDa		38–40
Glycine	48, 58 and 94 kDa		41, 42

[a] Abbreviations: DOC, deoxycholate; n-ACh, nicotinic acetylcholine; GABA, γ-aminobutyric acid.
[b] References for receptor sequence and purification.

Within each family, the receptors show homologies in their primary, secondary and quaternary structures, and similarities in their tertiary structures are also most probable (67, 70).

2. APPLICATION OF MOLECULAR GENETIC TECHNIQUES FOR THE CHARACTERIZATION OF RECEPTORS

Receptor amino acid sequence determinations have traditionally required initial isolation of the receptor membrane proteins. This has often involved laborious and troublesome methods for solubilization and purification, and the rate of progress in this field has been dependent on overcoming these difficulties. Recently the application of the powerful techniques of molecular genetics has altered this situation dramatically and the use of purified receptors is no longer necessary in order to establish receptor amino acid sequences. Instead, cDNA or genomic clones for receptors are isolated, and the receptor amino acid sequences then deduced using these clones. An important advantage is that abundant amounts of receptor can now be produced by these techniques for further use in studying cellular actions and mechanisms of signal transduction.

The complete amino acid sequence of a cell surface receptor for a peptide ligand was first published in 1984 and concerned the EGF receptor (161, 162). Initial purification of the receptor was by immunoaffinity chromatography, using monoclonal antibodies. As a consequence, a partial amino acid sequence was obtained which led to the design of a DNA hybridization probe. This was then used to deduce the complete receptor sequence by screening cDNA libraries derived from human placental tissue and epidermoid carcinoma cells, using the λ-phage vector system. In the following year, the human insulin receptor was purified and partially sequenced. This information was similarly used to design single, long, synthetic DNA probes for hybridization screening of a DNA library and identification of insulin receptor cDNA clones from which the amino acid sequence of the insulin receptor precursor was then deduced (163, 164).

Different receptors in the same group have some amino acid sequence similarity and it is possible to isolate cDNA or genomic clones for receptors in the same group, using oligonucleotide probes corresponding to the common sequences. The two different genomic

▶ p. 124

Cell Surface Receptors

Table 3. Classification of transmembrane receptors: examples of monomeric receptors with seven transmembrane segments functioning by activation of G-proteins transiently coupled to effector enzymes (solubilized using digitonin, digitonin/cholate mixtures, CHAPS, CHAPSO, dodecyl maltoside)[a]

Receptors for[a]	Subtype	Structural and functional characteristics and purification method	Refs[b]
Rhodopsin	—	Monomeric: chemical methods	43
Epinephrine (adrenaline)	α_1, α_2	64–66 kDa: affinity chromatography	44–48
Norepinephrine (noradrenaline)	β_1, β_2	62–65 kDa: affinity chromatography	49–51
mACh	M_1–M_5	70–90 kDa	52
A-II	—	Monomeric: expression of *mas* oncogene sequence	53
5-HT	1C	Monomeric: isolation of cDNA by + ve screening	54
LH/hCG	—	Probably monomeric (73–90 kDa): affinity chromatography	55–57

[a] Abbreviations: CHAPS, 3-(3-cholamidopropyl)dimethylammonio-1-propanesulfonate; CHAPSO, 3-(3-cholamidopropyl)dimethylammonio-2-hydroxy-1-propanesulfonate; mACh, muscarinic acetylcholine; A-II, angiotensin II; 5-HT, 5-hydroxytryptamine; LH/hCG, luteinizing hormone/human chorionic gonadotrophin; cDNA, complementary DNA.
[b] References for receptor sequence and purification.

Table 4. Receptors coupled to stimulation of phosphatidylinositol-4,5-bisphosphate hydrolysis

Receptor	Examples of hormonal or neurotransmitter actions[a]	Receptor subtype	Radioligand	Refs
Acetylcholine (muscarinic)	Fluid secretion in parotid	M_1	[^{3}H]quinuclidinyl benzilate	52, 71–74
	Enzyme secretion in exocrine pancreas	M_1	[^{3}H]pirenzepine	
	Hormone secretion in endocrine pancreas	M_3		
	Neuronal neurotransmitter functioning via muscarinic synapses			
α_1-Adrenoceptors	Glycogenolysis in liver	α_{1A}	[^{3}H]prazosine	60, 75, 76
	Vasoconstriction in smooth muscle	α_{1B}		
	Neuronal neurotransmitter functioning via α-adrenergic synapses	α_{1C}		
Angiotensin II	Aldosterone production by adrenal cortex regulates blood pressure–volume homeostasis	AT_1	[^{125}I]- or [^{3}H]angiotensin	60, 77, 78
Bombesin	Hormone secretion in endocrine pancreas	—	[^{125}I]Tyr-bombesin	79
	Direct and indirect effects on the gastrointestinal system			
Dopamine (D)	Activation of phospholipase C in renal tissue	D_1	—	60, 80–82
	Inositol trisphosphate formation and activation of Ca^{2+}-dependent Cl^- channels			
Histamine (H)	Catecholamine secretion by adrenal medulla	H_1	[^{3}H]mepyramine	60, 83, 84
5-Hydroxytryptamine (serotonin)	Fluid secretion in insect salivary gland	$5HT_2$	[^{3}H]ketanserin	85
	Role in cardiovascular regulation	$5HT_{1C}$	[^{3}H]5-hydroxytryptamine	60, 86–89
	Stimulates intestinal secretion			
	Neuronal neurotransmitter functioning via serotonin synapses			

Table 4. Continued

Receptor	Examples of hormonal or neurotransmitter actions[a]	Receptor subtype	Radioligand	Refs
Gonadotropin releasing hormone (GnRH, LHRH)	Stimulates release of LH and FSH from pituitary gonadotrophs	—	[^{125}I]LHRH	90, 91
Oxytocin	Role in corpus luteum function and lactating rat mammary gland Control of the estrous cycle in ruminants Role in human uterine myometrium and fetal membrane (amnion and chorion-decidua)	OT	[^{3}H]oxytocin	60, 92–95
Tachykinins (Substance P, K and neurokinin B)	Postsynaptic modulatory or regulatory peptides (e.g. in CNS and in pain sensory neurons of spinal chord) Act as neuroendocrine link within viscera (e.g. has diffuse constrictor activity in smooth muscle of gut, urogenital tract and lung)	NK_1 NK_2 NK_3	[^{3}H]- and [^{125}I]substance P	60, 96–100
Thrombin	Secretion of storage granules and aggregation of platelets	—	—	101–103
Thyrotropin releasing hormone (TRH)	Stimulation of prolactin and thyroid stimulating hormone secretion from pituitary	—	[^{3}H](3 methyl His)-TRH	104–106
Vasopressin (V)	Promotion of water retention in kidney tubule membrane Vasoconstriction of smooth muscle Potentiates CRF-stimulated release of ACTH from corticotrophs	V_{1A} V_{1B}	[^{3}H]vasopressin	60, 78, 106, 107

[a] Abbreviations: LH, luteinizing hormone; FSH, follicle stimulating hormone; CNS, central nervous system; CRF, corticotropin releasing factor; ACTH, adrenocorticotropic hormone (corticotropin).

Table 5. Receptors coupled to adenylate cyclase (stimulation of cyclic AMP production)

Receptor	Examples of hormonal or neurotransmitter actions	Receptor subtype	Radioligand	Refs
β-Adrenoceptors	Neuronal neurotransmitter functioning via β-adrenergic synapses	β_1; β_2 β_3	[^{3}H]dihydroalprenolol [^{125}I]iodocyanopindolol	49–51, 60, 108–110
γ-Aminobutyric acid (GABA)	Neuronal neurotransmitter functioning via GABAergic synapses	$GABA_B$	[^{3}H]baclofen	38–40, 111–112
Corticotropin (ACTH)	Stimulates corticosteroidogenesis in adrenal cortex	—	[^{125}I]ACTH	113–115
Corticotropin releasing factor (CRF)	Stimulates ACTH and β-endorphin release from pituitary corticotrophs	—	[^{125}I]CRF	106, 116–118
Dopamine	Neuronal neurotransmitter functioning via dopaminergic synapses	D_1 D_5	[^{3}H] SCH 23390 [^{125}I] SCH 23982	119–123
Follicle stimulating hormone (FSH)	Stimulates preovulatory ovarian estrogen production and follicle maturation Controls differentiation and activity of the Sertoli cells in seminiferous tubules of the testis	—	[^{125}I]FSH	124, 125
Glucagon	Induces glycogenolysis, gluconeogenesis and ketogenesis in liver Also affects heart, endocrine pancreas and adipose tissue	—	[^{125}I]glucagon	126, 127
Growth hormone releasing hormone (GRH) and Somatostatin (SRIF)	Stimulation (GRH) and inhibition (SRIF) of growth hormone release by pituitary somatotrophs	—	[^{125}I]GRH [^{125}I]Tyr-SRIF	106, 128–130

Table 5. Continued

Receptor	Examples of hormonal or neurotransmitter actions	Receptor subtype	Radioligand	Refs
Histamine	Neuronal neurotransmitter functioning via histaminergic synapses Stimulates gastric secretion in the gastrointestinal tract	H_2	[^{3}H]tiotidine; molecular cloned gene for H_2 receptor	60, 131–135
Luteinizing hormone (LH)	Stimulates androgen production in testicular Leydig cells Controls (together with FSH) differentiation of the follicles, ovulation and ovarian steroidogenesis	—	[^{125}I]LH [^{125}]hCG[a]	55–57, 135
Oxytocin	Promotes contraction of smooth muscle cells in the uterus and mammary glands Stimulates milk ejection by myoepithelial cells around nipple of mammary gland	V_2	[^{3}H]oxytocin [^{3}H]deamino-[Arg^8]VP	60, 136, 137
Prostaglandins (PG)	Produced by most tissues and implicated in a wide range of physiological processes; probably act in a paracrine manner on various tissues by modulating hormonal and neurohormonal signals	DP IP EP_2	[^{3}H]PGD_2 [^{3}H]iloprost [^{3}H]PGE_2	60, 138–142
Vasoactive intestinal peptide (VIP)	Stimulates prolactin and GH release Induces vasodilation in vascular systems and increases heart rate	—	[^{125}I]VIP	106, 143–145
Vasopressin	Promotes retention of water in kidney tubule membrane Acts on vascular smooth muscle cells, medullary interstitial cells and mesangial cells to promote prostacyclin or PGE_2 production, in addition to a contractable response	V_2	[^{3}H]vasopressin	60, 158–160

[a] hCG = human chorionic gonadotropin.

Table 6. Receptors coupled to adenylate cyclase (inhibition of cyclic AMP production)

Receptor	Examples of hormonal or neurotransmitter actions	Receptor subtype	Radioligand	Refs
Acetylcholine (muscarinic)	Very diverse, including effects on exocrine secretory glands (e.g. paratoid, pancreas), smooth muscle (e.g. stomach, iris), brain and nervous system	M_2 M_4	[^{3}H]methyl-scopolamine [^{3}H]quinuclidinyl benzilate	52, 71, 146, 147
α_2-Adrenoceptors	Inhibitory role in insulin secretion in pancreatic B-cell Diverse controls on the state of the α-adrenergic synapses Role in 5-hydroxytryptamine release	α_2 subtypes: $\alpha_{2A}, \alpha_{2B}, \alpha_{2C}$	[^{3}H]RX781094	48, 148–152
Dopamine	Inhibits prolactin release from pituitary lactotrophs	D_2	[^{3}H]spiperone	60, 106, 152–154
5-Hydroxy-tryptamine (serotonin)	Neuronal transmitter functioning via serotonin synapses	$5HT_{1A}$ $5HT_{1D}$	[^{3}H]ipsapirone [^{125}I]5-HT-5-O-carboxymethyl-glycyltyrosin-amide	60
Opioid	Reduces the excitability and impulse firing rate of the postsynaptic cell in the brain	μ δ	[^{3}H]-D-ala^2, D-leu^5-enkephalin D-pen^2, D-pen^5 tyrosyl-[^{3}H]-enkephalin	60, 155–157

clones isolated by this approach, using partial sequences of a subtype of the mACh receptors, were shown to be genes for previously unidentified subtypes of mACh receptors (165). As the number of receptors with known sequences multiply, this strategy has been used more often, and by this means the sequences of dopamine D_2 and variants of β_1- and β_2-adrenergic receptors have been determined (166–168, see also ref. 60). Sequences of the $5HT_{1C}$ and substance K (neurokinin) receptors were deduced from cDNA clones that had been isolated using functional assays without knowing the partial amino acid sequences (161, 169, 170). Receptor activities were followed by assessing changes in membrane potentials induced in *Xenopus* oocytes, injected with mRNAs transcribed from successively fractionated cDNA clones. In this strategy (positive selection) the length of the mRNA may be a limiting factor, since the cDNA clone must contain the full length that is necessary for expression of the receptor activity. Another approach (negative selection) has been used for isolation of cDNA clones for 5-HT receptors, and was also based on electrophysiological assays of oocytes injected with mRNA. After single-stranded DNA derived from a cDNA library had been used to deplete the mRNA encoding the receptor, the reduced expression of receptor activity was then assessed (171). Both positive and negative selection techniques will continue to be of use for the characterization of receptors.

Recently the polymerase chain reaction (PCR) was used to amplify a cDNA sequence selectively. This cDNA fragment was then used as a probe to isolate a cDNA clone from a genomic library for the characterization of receptors (e.g. for the histamine H_2 and the dopamine D_1 receptors) (83, 133, 172).

Over the past few years there have been rapid strides in this field. For example, there are likely to be over 150 different receptors that are coupled to G-proteins, and more than 40 of these have now been cloned (134). Presumably, the remainder will be cloned in the foreseeable future, by taking advantage of the above molecular biological approaches. Other powerful techniques of benefit to the field include site-directed mutagenesis (to produce hormone variants), the production of chimeric proteins (from chimeric gene constructs), the transfection of cells with DNA encoding the receptor and expression studies (to determine the time course and levels of receptor mRNA expression).

3. KINETIC BINDING CONSTANTS FOR HORMONE–RECEPTOR INTERACTIONS

The binding constants obtained using static incubation methodology for a variety of receptors are given in *Table 7*. For this representative sample the equilibrium constant (K_D) is seen to vary between 10^{-8} and 10^{-11} M for low- and high-affinity receptors, respectively. In these evaluations, little attempt was made to correct for degradation of the radioligand during the incubations, and consequently K_D values determined in static systems are usually higher than the concentrations found to give half-maximal biological responses (i.e. the ED_{50} value). For example, for ACTH the K_D value is 1–3 orders of magnitude higher than the ED_{50} value, which is 4×10^{-12} M with adrenal fasciculata cells (173).

Kinetic constants have also been determined using flowing-system methodology (173), and the dynamic constants have been obtained for ACTH binding to whole isolated adrenal fasciculata cells. The K_D values thereby obtained with these whole cells closely approached adrenal fasciculata cell ED_{50} values for steroidogenesis (see *Table 7*) and the method allowed k_{-1} and k_{+1} evaluations.

4. GENERAL CONSIDERATIONS

The number of new hormones, growth factors and neurotransmitters is expanding rapidly and, as a consequence, the cell surface receptor field is continually increasing. The attempt

▶ p. 131

Table 7. Hormone binding constants obtained by static methodology

Hormone or ligand	Tissue	Temp. (°C)	Binding sites[a]	K_D (M)	k_{-1} (min^{-1})	k_{+1} (nM^{-1} min^{-1})	Ref.
Mono[^{125}I]glucagon	Rat liver membrane	32.5	I	7.5×10^{-10}			174
[^{125}I]Human chorionic gonadotropin (hCG) and [^{125}I]-lutropin (LH)	Dissociated rat luteal cells	37 37	I (hCG) I (LH)	1.8×10^{-10} 2.6×10^{-10}	0.02 0.015[b]	0.09 0.06	175
	Quail testicular plasma membrane fraction		I II	3×10^{-9} 3×10^{-8}			176
	Affinity purified gonadotropin receptor from rat ovary		I	2.5×10^{-9}			177
[^{125}I]Follicle stimulating hormone (FSH)	Cultured pig granulosa cells isolated from small follicles		Normal (cultured + EGF)	2.5×10^{-10} 5.3×10^{-10}			178
[^{3}H]Prostaglandin E_1	Bovine luteal membranes and dissociated cells	38	II	1.3×10^{-9}	0.17	0.011	179
			II	2.4×10^{-9}	0.23	0.025	
	Rabbit gastric mucosa		I	5.3×10^{-9}			180
[^{125}I]Adreno-corticotropin (ACTH)	Dissociated rat adrenal fasciculata cells	37	I II	2.5×10^{-10} 1×10^{-8}			181
	Dissociated rat adrenocortical cells		I II	2.6×10^{-10} 7.1×10^{-9}			182

Table 7. Continued

Hormone or ligand	Tissue	Temp. (°C)	Binding sites[a]	K_D (M)	k_{-1} (min^{-1})	k_{+1} ($nM^{-1} min^{-1}$)	Ref.
	Dissociated rat adrenocortical cells	23	I	1.4×10^{-9}	[c]		183
	Guinea-pig adrenal cortex slices		I	6×10^{-10}			184
	Rat adrenal glomerulosa cells	22	I II	7.6×10^{-11} 1.2×10^{-9}			185
	Rat adrenal fasciculata cells	22	I II	1.1×10^{-11} 2.9×10^{-9}			
	Immunoaffinity purified ACTH receptor from Y-I cells	4	I II	2.9×10^{-11} 1×10^{-9}			186
	Dissociated rat adrenal fasciculata cells[d]	22	I	4.6×10^{-11}	0.23	5	173
[^{125}I]Hydroxy-benzylpindolol: a β-adrenergic antagonist	Turkey erythrocyte membranes	20 37	I I	2.6×10^{-11} 1×10^{-11}	0.08 0.03	6 54	187
β_2-Adrenoceptors ([^{125}I]CYP)[e]	Guinea-pig alveolar macrophages		I	2.4×10^{-11}			188
β-Adrenoceptor subtypes ([^{125}I]CYP)[e]	Rat cerebral cortex membranes		I High and low affinity states[f]	2×10^{-11}			189[f]

Table 7. Continued

Hormone or ligand	Tissue	Temp. (°C)	Binding sites[a]	K_D (M)	k_{-1} (min^{-1})	k_{+1} ($nM^{-1}\ min^{-1}$)	Ref.
$[^{125}I]$Somatostatin (SRIF)	Chicken pituitary glands (crude membranes)		I	1×10^{-9}			190
$[^{125}I]$Human growth hormone releasing hormone (GRH)	Dissociated rat anterior pituitary cells	24	I	$2\text{–}7 \times 10^{-10}$			191
	Pituitary adenoma cell membranes from acromegalic patients	24	I	3×10^{-10}			192
$[^{125}I]$Angiotensin II (AII)	Dissociated rat and dog glomerulosa cells and adrenal cortex membrane	37 37	I (rat) I (dog) I (mem.)	3.7×10^{-10} 9×10^{-10} 5×10^{-10}			193
	Rat kidney epithelial membrane		I	6.2×10^{-10}			193
$[^{125}I]$Angiotensin II	Ovine and bovine adrenal fasciculata cells	37	I	2×10^{-9}			193
$[^{3}H]$Angiotensin II	Basolateral membrane from rat proximal renal tubule		I	2.3×10^{-10}			194

Table 7. Continued

Hormone or ligand	Tissue	Temp. (°C)	Binding sites[a]	K_D (M)	k_{-1} (min^{-1})	k_{+1} ($nM^{-1} min^{-1}$)	Ref.
[125I]Nerve growth factor (NGF)	Dissociated cells from chick embryonic dorsal root ganglia (day 8)	37	I II	2.3×10^{-11} 1.7×10^{-9}			195, 196
	Dissociated cells from chick embryonic sympathetic ganglia (day 9)		I II	3×10^{-11} 1.8×10^{-9}			197
	PC12 pheochromocytoma cell line	37	I II	2×10^{-10} 2×10^{-10}	1.4 0.03	7[b] 0.15[b]	197, 198
NGF–α-macroglobulin complex	Injected intravenously in mouse or combined *in vitro* with plasma		I II	2.9×10^{-9} 1.2×10^{-6}			199
[125I]Epidermal growth factor (EGF; urogastrone)	Human fibroblasts in monolayer culture	4	Not a simple, one-step process. Satisfactory fits given by ternary complex model		0.3	72	200
	Normal and neoplastic human endometrium		I	6.4×10^{-10}			201

Table 7. Continued

Hormone or ligand	Tissue	Temp. (°C)	Binding sites[a]	K_D (M)	k_{-1} (min^{-1})	k_{+1} (nM^{-1} min^{-1})	Ref.
	Rat mammary cells			1.7×10^{-10} 2.7×10^{-10}	(Phosphatidyl-ethanolamine deficient cells)		202
[^{125}I]Platelet-derived growth factor (PDGF)	Swiss 3T3 cell line Monkey smooth muscle cells	4 4	I (hook effect)	$0.7–1.7 \times 10^{-11}$			203
Vasculotropin (structurally related to PDGF)	Chinese hamster ovary cells, using an expression vector		I	4×10^{-10}			204
	Bovine adrenal cortex-derived capillary endothelial cells		I II	2×10^{-12} 0.8×10^{-10}			205
[^{125}I]Insulin-like growth factor-I (IGF-I)	Normal and neoplastic human endometrial membranes			6×10^{-9}			206
	Bovine adrenal fasciculata cell cultures		(IGF-I R) (Insulin R)	1.4×10^{-9} 1×10^{-9}			207
	Granulosa cells from normal (N) and FSH-treated (F) rats		(N rats) (F rats)	1.9×10^{-9} 2.6×10^{-9}			208

Table 7. Continued

Hormone or ligand	Tissue	Temp. (°C)	Binding sites[a]	K_D (M)	k_{-1} (min^{-1})	k_{+1} (nM^{-1} min^{-1})	Ref.
Interleukin-3 (IL-3)	Human basophilic granulocytes			2×10^{-10}			209
IL-3 and granulocyte–macrophage colony-stimulating factor (GM-CSF)	Human eosinophils (e) and neutrophils (n)	4	e + IL-3	4.7×10^{-10}			210
			n + IL-3	No binding			
			e + GMCS	4.4×10^{-11}			
			n + GMCS	7×10^{-11}			
[^{125}I]Growth hormone (GH)	Cultured rat epiphyseal chondrocytes	Max. binding 24	I	4.6×10^{-10}			211

[a] High affinity or 'fast receptor' (site I); low affinity or 'slow receptor' (site II) (198).
[b] Not reported but calculated from published data.
[c] Rate constant for dissociation given, but in the units of the rate constant for association (i.e. 4×10^{-3} M^{-1} min^{-1}) (183).
[d] Dynamic studies using a superfusion system (173).
[e] Iodocyanopindalol.
[f] Also binds to serotonin receptors (189).

here has been to provide representative examples and to portray receptor groupings and superfamilies. Valuable publications concern receptor methodology (67, 212), α_2-adrenergic receptors (213), β-adrenergic receptors (214), muscarinic receptors (215), serotonin receptors (216), opiate receptors (217), growth factors (218) and LDL receptors (24).

The molecular biological approaches outlined will greatly assist the characterization of the receptors. For example, *in situ* hybridization techniques will localize yet to be identified receptor sites and iterative site-directed mutagenesis will be used to engineer hormone variants (219) to aid the design and understanding of protein–protein interfaces. As the incomplete picture of receptor varieties and mechanisms becomes more complete, it is to be expected that a clearer picture of the superfamily relationships will emerge.

5. REFERENCES

1. Hucho, F. (1986) *Neurochemistry, Fundamentals and Concepts*. VCH, Weinheim.

2. Ebina, Y., Ellis, L., Jarnagin, K., Edery, M., Graf, L., Clauser, E., Ou, J.-H., Masiarz, F., Kan, Y.W., Goldfine, I.D., Roth, R.A. and Rutter, W.J. (1985) *Cell*, **40**, 747.

3. Ullrich, A., Bell, J.R., Chen, E.Y., Herrera, R., Petruzelli, L. M., Dull, T.J., Gray, A., Coussens, L., Liao, Y.-C., Tsubokawa, M., Mason, A., Seeburg, P.H., Grunfeld, C., Rosen, O.M. and Ramachandran, J. (1985) *Nature*, **313**, 756.

4. Roth, R.A. and Cassell, D.J. (1983) *Science*, **219**, 299.

5. Fujita-Yamaguchi, Y., Choi, Y., Sakamoto, Y. and Itakura, K. (1983) *J. Biol. Chem.*, **258**, 5045.

6. Goldfine, I.D. (1987) *Endocrin. Revs*, **8**, 235.

7. Ullrich, A., Gray, A., Tam, A.W., Yang-Feng, T., Tsubokawa, M., Collins, C., Henzel, W., LeBon, T., Kathuria, S., Chen, E., Jacobs, S., Francke, U., Ramachandran, J. and Fujita-Yamaguchi, Y. (1986) *EMBO J.*, **5**, 2053.

8. LeBon, T.R., Jacobs, S., Cuatrecasas, P., Kathuria, S. and Fujita-Yamaguchi, Y. (1986) *J. Biol. Chem.*, **261**, 7685.

9. Ullrich, A. and Schlessinger, J. (1990) *Cell*, **61**, 203.

10. Yarden, Y., Escobedo, J.A., Kuang, W.J., Yang-Feng, T.L., Daniel, T.O., Tremble, P.M., Chen, E.Y., Ando, M.E., Harkins, R.N., Francke, U., Fried, V.A., Ullrich, A. and Williams, L.T. (1986) *Nature*, **323**, 226.

11. Morgan, D.O., Edman, J.C., Standring, D.N., Fried, V.A., Smith, M.C., Roth, R.A. and Rutter, W.J. (1987) *Nature*, **329**, 310.

12. Nishimoto, I., Murayama, Y. and Okamoto, T. (1991) in *Modern Concepts of IGFs* (E.M. Spencer ed.). Elsevier Science, Amsterdam, p. 517.

13. Radeke, M.J., Misko, T.P., Hsu, C., Herzenberg, L.A. and Shooter, E.M. (1987) *Nature*, **325**, 593.

14. Johnson, D., Lanahan, A., Buck, C.R., Sehgal, A., Morgan, C., Mercer, E., Bothwell, M. and Chao, M.V. (1986) *Cell*, **47**, 545.

15. Welcher, A.A., Bitter, C.M., Radeke, M.J. and Shooter, E.M. (1991) *Proc. Natl. Acad. Sci.*, **88**, 159.

16. Ullrich, A., Coussens, L., Hayflick, J.S., Dull, T.J., Gray, A., Tam, A.W., Lee, J.,Yarden, Y., Liberman, T.A., Schlessinger, J., Downward, J., Mayes, E.L.V., Whittle, N., Waterfield, M.D. and Seeburg, J. (1984) *Nature*, **309**, 418.

17. Panayotou, G.N. and Gregoriou, M. (1990) in *Receptor Biochemistry: A Practical Approach* (E.C. Hulme ed.). IRL Press, Oxford University Press, Oxford, p. 203.

18. Daniel, T., Tremble, P.M., Frackelton, A.R. and Williams, L.T. (1985). *Proc. Natl. Acad. Sci. USA*, **82**, 2684.

19. Hart, C.E., Forstrom, J.W., Kelly, J.D., Seifert, R.A., Smith, R.A., Ross, R., Murray, M.J. and Bowen-Pope, D.F. (1988) *Science*, **240**, 1529.

20. Heldin, C.-H., Backstrom, G., Ostman, A., *et al.* (1988) *EMBO J.*, **7**, 1387.

21. Westermark, B. (1990) *Acta Endocrinol.*, **123**, 131.

22. Takayanagi, R., Inagami, T., Snajdar, R.M., Imada, T., Tamura, M. and Misono, K.S. (1987) *J. Biol. Chem.*, **262**, 12 104.

23. Chinkers, M., Garbers, D.L., Chang, M.-S., Lowe, D.G., Chin, H., Goeddel, D.V. and Schultz, S. (1989) *Nature*, **338**, 78.

24. Myant, N.B. ed. (1990) *Cholesterol Metabolism, LDL and the LDL Receptor*. Academic Press, London.

25. Goldstein, J.L., Brown, M.S., Anderson, R.G., Russel, D.W. and Schneider, W.J. (1985) in *Annual Review of Cell Biology* (G.E. Palade, ed.). Annual Reviews Inc., Palo Alto, CA, Vol. 1, p. 1.

26. Robenek, H., Harrach, B. and Severs, N.J. (1991) *Arterioscler. Thromb.*, **11**, 261.

27. Leonard, W.J., Depper, J.M., Crabtree, G.R., Rudikoff, S., Pumphrey, J., Robb, R.J., Kronke, M., Svetlik, P.B., Peffer, N.J., Waldmann, T.A. and Greene, W.C. (1984) *Nature*, **311**, 626.

28. Nikaido, T., Shimizu, A., Ishida, N., Sabe, H., Teshigawara, K., Maeda, M., Uchiyama, T., Yodoi, T. and Honjo, T. (1984) *Nature*, **311**, 631.

29. Guy, G.R., Bee, N.S. and Peng, C.S. (1990) *Progress in Growth Factor Research*, **2**, 45.

30. Gospodarowicz, D., Ferrara, N., Schweigerer, L. and Neufeld, G. (1987) *Endocrin. Revs.*, **8**, 95.

31. Leung, D.W., Spencer, S.A., Cachianes, G., Hammonds, R.G., Collins, C., Henzel, W.J., Barnard, R., Waters, M.J. and Wood, W.I. (1987) *Nature*, **330**, 537.

32. Boutin, J.-M., Jolicoeur, C., Okamura, H., Gagnon, J., Edery, M., Shirota, M., Banville, D., Dusanter-Fourt, I., Dijane, J. and Kelly, P.A. (1988) *Cell*, **53**, 69.

33. Spencer, S.A., Hammonds, R.D., Henzel, W.J., Rodriguez, H., Waters, M.J. and Wood, W.I. (1988) *J. Biol. Chem.*, **263**, 7862.

34. Spencer, S.A., Leung, D.W., Godowski, P.J., Hammonds, R.G., Waters, M.J. and Wood, W.I. (1990) *Rec. Prog. Horm. Res.*, **46**, 165.

35. Noda, M., Takahashi, H., Tanabe, T., Toyosato, M., Furutani, Y., Hirose, T., Asai, M., Inayama, S., Miyata, T. and Numa, S. (1982) *Nature*, **299**, 793.

36. Wu, W.C.-S. and Raftery, M.A. (1981) *Biochemistry*, **20**, 694.

37. Hertling-Jaweed, S., Bandini, G. and Hucho, F. (1990) in *Receptor Biochemistry: A Practical Approach* (E.C. Hulme ed.). IRL Press, Oxford University Press, Oxford, p. 163.

38. Schofield, P.R., Darlison, M.G., Fujita, N., Burt, D.R., Stephenson, F.A., Rodriguez, H., Rhee, L. M., Ramachandran, J., Reale, V., Glencorse, T.A., Seeburg, P.H. and Barnard, E.A. (1987) *Nature*, **328**, 221.

39. Siegel, E. and Barnard, E.A. (1984) *J. Biol. Chem.*, **259**, 7219.

40. Stephenson, F.A. (1990) in *Receptor Biochemistry: A Practical Approach* (E.C. Hulme ed.). IRL Press, Oxford University Press, Oxford, p. 177.

41. Grenningloh, G., Rienitz, A., Schmitt, B., Methfessel, C., Zensen, M., Beyreuther, K., Gundelfinger, E.D. and Betz, H. (1987) *Nature*, **328**, 215.

42. Graham, D., Pfeiffer, F., Simler, R. and Betz, H. (1985) *Biochemistry*, **24**, 990.

43. Ovchinnikov, Y.A. (1982) *FEBS Lett.*, **148**, 179.

44. Regan, J.W., Nakata, H., DeMarinis, R.M., Caron, M.G. and Lefkowitz, R.J. (1986) *J. Biol. Chem.*, **261**, 3894.

45. Lomasney, J.W., Leeb-Lundberg, L.M., Cotecchia, S., Regan, J.W., DeBernardis, J.F., Caron, M.G. and Lefkowitz, R.J. (1986) *J. Biol. Chem.*, **261**, 7710.

46. Kobilka, B.K., Matsui, H., Kobilka, T.S., Yang-Feng, T.L., Francke, U., Caron, M.G., Lefkowitz, R.J. and Regan, J.W. (1987) *Science*, **238**, 650.

47. Cotecchia, S., Schwinn, D.A., Randall, R.R., Lefkowitz, R.J., Caron, M.G. and Kobilka, B.K. (1989) *Proc. Natl. Acad. Sci. USA*, **85**, 7159.

48. Regan, J.W. and Matsui, H. (1990) in *Receptor Biochemistry: A Practical Approach* (E.C. Hulme ed.). IRL Press, Oxford University Press, Oxford, p. 141.

49. Benovic, J.L., Shorr, R.G.L., Caron, M.G. and Lefkowitz, R.J. (1984) *Biochemistry*, **23**, 4510.

50. Dixon, R.A.F., Kobilka, B.K., Strader, D.J., Benovic, J.L., Dohlman, H.J., Frielle, T., Bolanowski, M.A., Bennett, C.D., Rands, E., Diehl, R.E., Mumford, R.A., Slater, E.E., Siegel, I.S., Caron,, M.G., Lefkowitz, R.J. and Strader, C.D. (1986) *Nature*, **321**, 75.

51. Yarden, Y., Rodriguez, H., Wong, S.K.-F., Brand, D.R., May, D.C., Burnier, J., Harkins, R.N., Chen, E.Y., Ramachandran, J., Ullrich, A. and Ross, E.M. (1986) *Proc. Natl. Acad. Sci. USA*, **83**, 6795.

52. Haga, T., Haga, K. and Hulme E.C. (1990) in *Receptor Biochemistry: A Practical Approach* (E.C. Hulme ed.). IRL Press, Oxford University Press, Oxford, p. 51.

53. Jackson, T.R., Blair, A.C., Marshall, J., Goedert, M. and Hanley, M.R. (1988) *Nature*, **335**, 437.

54. Julius, D., Dermott, A.B., Axel, R. and Jessell, T.M. (1988) *Science*, **241**, 558.

55. McFarland, K.C., Sprengel, R., Philips, H.S., Kohler, M., Rosemblit, N., Nikolics, K., Segaloff, D.L. and Seeburg, P.H. (1989) *Science*, **245**, 494.

56. Ascoli, M. and Segaloff, D.L. (1989) *Endocrin. Revs*, **10**, 27.

57. Segaloff, D.L., Sprengel, R., Nikolics, K. and Ascoli, M. (1990) *Rec. Prog. Horm. Res.*, **46**, 261.

58. Dohlman, H.G., Caron, M.G. and Lefkowitz, R.J. (1987) *Biochemistry*, **26**, 2657.

59. Harrison, J.K., Pearson, W.R. and Lynch, K.R. (1991) *Trends in Pharm. Sci.*, **12**, 62.

60. Andersen, P.H., Gingrich, J.A., Bates, M.D., Dearry, A., Falardeau, P., Senogles, S.E. and Caron, M.G. (1990) *Trends in Pharm. Sci.*, **11**, 231; and Sibley, D.R. (1991) *Trends in Pharm. Sci.*, **12**, 7; and *Receptor Nomenclature Suppl.*

61. Birdsall, N.J.M. (1991) *Trends in Pharm. Sci.*, **12**, 9.

62. Levitzki, A. (1980) in *Cellular Receptors for Hormones and Neurotransmitters* (D. Schulster and A. Levitzki eds). Wiley and Sons, Chichester, p. 9.

63. Iyengar, R., Birnbaumer, L., Schulster, D., Houslay, M. and Michell, R.H. (1980) in *Cellular Receptors for Hormones and Neurotransmitters* (D. Schulster and A. Levitzki eds). Wiley and Sons, Chichester, p. 55.

64. Changeux, J.-P., Devillers-Thiery, A. and Chemouille, P. (1984) *Science*, **225**, 1335.

65. Williams, A.F. and Barelay, A.N. (1988) *Ann. Rev. Immunol*, **6**, 381.

66. Cosman, D., Lyman, S.D., Idzerda, R.L., Beckmann, N.P., Park, L.S., Goodwin, R.G. and March, C.F. (1990) *Trends in Bioch. Sci.*, **15**, 265.

67. Hulme, E.C. ed. (1990) *Receptor Biochemistry: A Practical Approach.* IRL Press, Oxford University Press, Oxford.

68. Schulster, D. (1988) in *Radiochemicals in Biomedical Research* (E.A. Evans and K.G. Oldham eds). Wiley and Sons, Chichester, p. 94.

69. Strange, P.G. (1988) in *Radiochemicals in Biomedical Research* (E.A. Evans and K.G. Oldham eds). Wiley and Sons, Chichester, p. 56.

70. Green, N.M. (1990) in *Receptor Biochemistry: A Practical Approach.* IRL Press, Oxford University Press, Oxford, p. 277.

71. Yamamura, H.I. and Snyder, S.H. (1974) *Proc. Natl. Acad. Sci., USA*, **71**, 1725.

72. Masters, S.B., Harden, T.K. and Brown, J.H. (1984) *Mol. Pharmacol.*, **26**, 149.

73. Watson, M., Yamamura, H.I. and Roeske, W.R. (1983) *Life Sci.*, **32**, 3001.

74. Brown, E., Kendall, D.A. and Nahorski, S.R. (1984) *J. Neurochem.*, **42**, 1379.

75. Greengrass, P. and Bremner, R. (1979) *Eur. J. Pharmacol.*, **55**, 323.

76. Villalobos-Molina, R., Mirna, U.C., Hong, E. and Garcia-Sainz, J.A. (1982) *J. Pharmacol. Exp. Ther.*, **222**, 258.

77. Glossmann, H., Baukul, A.J. and Catt, K.J. (1974) *J. Biol. Chem.*, **249**, 825.

78. Creba, J.A., Downes, C.P., Hawkins, P.T., Brewster, G., Michell, R.H. and Kirk, C.J. (1983) *Biochem J.*, **212**, 733.

79. Gardner, J.D. and Jensen, R.T. (1982) in *Hormone Receptors* (L.D. Kohn ed.). Wiley and Sons, Chichester, Vol. 6, p. 277.

80. Felder, C.C., Jose, P.A. and Axelrod, J. (1989) *J. Pharmacol. Exp. Ther.*, **248**, 171.

81. Felder, C.C., Blecher, M. and Jose, P.A. (1989) *J. Biol. Chem.*, **264**, 8739.

82. Mahan, L.C., Burch, R.M., Monsma, F.J., Jr. and Sibley, D.R. (1990) *Proc. Natl. Acad. Sci. USA*, **87**, 2196.

83. Monsma, F.J., Mahan, L.C., McVittie, L.D., Gerfen, C.R. and Sibley, D.R. (1990) *Proc. Natl. Acad. Sci. USA*, **87**, 6723.

84. Daum, P.R., Downes, C.P. and Young, J.M. (1983) *Eur. J. Pharmacol.*, **87**, 497.

85. Leysen, J.E., Niemegeers, C.J.E., van Nueten, J.M. and Laduron, P.M. (1982) *Mol. Pharmacol.*, **21**, 301.

86. Mate, L., Poston, G.J. and Thompson, J.C. (1987) in *Gastrointestinal Endocrinology* (J.C. Thompson, G.H. Greeley, P.L. Rayford and C.M. Townsend eds). McGraw Hill, New York, p. 365.

87. Middlemiss, D.N. and Fozard, J.R. (1983) *Eur. J. Pharmacol.*, **90**, 151.

88. Morgan, D.G., May, P.C. and Finch, C.E. (1988) in *Receptors and Ligands in Neurological Disorders* (A.K. Sen and T.Lee eds). Cambridge University Press, Cambridge, p. 120.

89. Saxena, P.R. and Villalon, C.M. (1990) *Trends in Pharm. Sci.*, **11**, 95.

90. Naor, Z. (1990) *Endocrin. Revs*, **11**, 326.

91. Jennes, L. and Conn, P.M. (1988) in *Hormones and their Actions, Part II* (B.A. Cooke, R.J.B. King and H.J. van der Molen eds). Elsevier Science, Amsterdam, p. 135.

92. Flint, A.P.F., Sheldrick, E.L., Jones, D.S.C. and Auletta, F.J. (1989) *J. Reprod. Fert. Suppl.*, **37**, 195.

93. Wallace, J.M., Morgan, P.J., Helliwell, R., Aitken, R.P., Cheyne, M. and Williams, L.M. (1991) *J. Endocrinol.*, **128**, 187.

94. Pettibone, D.J., Woyden, C. and Totara, J.A. (1990) *Eur. J. Pharmacol.*, **188**, 235.

95. Benedetto, M.T., De Cisso, F., Rossiello, F., Nicosia, A.L., Lupi, G., Dell'Aqua, S. (1990) *J. Steroid Biochem.*, **35**, 205.

96. Jessel, T.M. and Womack, M.D. (1985) *Trends Neurochem. Sci.*, **8**, 43.

97. Mantyh, P.W., Pinnock, R.D., Downes, C.P., Goedert, M. and Hunt, S.P. (1984) *Nature*, **309**, 795.

98. Walker, J.P. and Thompson, J.C. (1987) in *Gastrointestinal Endocrinology* (J.C. Thompson, G.H. Greeley, P.L. Rayford and C.M. Townsend eds). McGraw Hill, New York, p. 317.

99. Regoli, D., Drapeau, G., Dion, S. and Couture, R. (1988) *Trends in Pharm. Sci.*, **9**, 290.

100. Mousli, M., Bueb, J.-L., Bronner, C., Rouot, B. and Landry, Y. (1990) *Trends in Pharm. Sci.*, **11**, 358.

101. Rittenhouse, S.E. (1985) in *Inositol and Phosphoinositides: Metabolism and Regulation* (J.E. Bleasdale, J. Eichberg and G. Hauser eds). Humana Press, Clifton, New Jersey, p. 459.

102. Lapetina, E.G. (1985) in *Inositol and Phosphoinositides: Metabolism and Regulation* (J.E. Bleasdale, J. Eichberg and G. Hauser eds). Humana Press, Clifton, New Jersey, p. 475.

103. Broekman, M.J. (1985) in *Inositol and Phosphoinositides: Metabolism and Regulation* (J.E. Bleasdale, J. Eichberg and G. Hauser eds). Humana Press, Clifton, New Jersey, p. 529.

104. Taylor, R.L. and Burt, D.R. (1982) *J. Neurochem.*, **38**, 1649.

105. Drummond, A.H., Bushfield, M. and MacPhee, C.H. (1984) *Mol. Pharmacol.*, **25**, 201.

106. Denef, C. (1988) in *Hormones and their Actions, Part II* (B.A. Cooke, R.J.B. King and H.J. van der Molen eds). Elsevier, Amsterdam, p. 113.

107. Gaillard, R.C., Schoenenberg, P., Farrod-Coune, C.A., Muller, A.F., Marie, J., Bockaert, J. and Jard, S. (1984) *Proc. Natl. Acad. Sci. USA*, **81**, 2907.

108. Lefkowitz, R.J., Limbird, L.E., Mukherjee, C. and Caron, M.G. (1976) *Biochim. Biophys. Acta*, **457**, 1.

109. Bylund, D.B. and Snyder, S.H. (1976) *Mol. Pharmacol.*, **12**, 568.

110. Benovic, J.L. (1990) in *Receptor Biochemistry: A Practical Approach* (E.C. Hulme ed.). IRL Press, Oxford UniversityPress, Oxford, p.125.

111. Hill, D.R. and Bowery, N.G. (1981) *Nature*, **290**, 149.

112. Hill, D.R., Bowery, N.G. and Hudson, A.L. (1984) *J. Neurochem.*, **42**, 652.

113. Ramachandran, F. (1987) in *Hormonal Proteins and Peptides* (C.H. Li ed.). Academic Press, New York, p. 31.

114. Schulster, D. and Schwyzer, R. (1980) in *Cellular Receptors* (D. Schulster and A. Levitzki eds). Wiley and Sons, Chichester, p. 197.

115. Hornsby, P.J. (1988) in *Hormones and their Actions Part II* (B.A. Cooke, R.J.B. King and H.J. van der Molen eds). Elsevier, Amsterdam, p. 193.

116. Litvin, Y., Pasmantier, R., Fleischer, N. and Erlichman, J. (1984) *J. Biol. Chem.*, **259**, 10 296.

117. Axelrod, J. and Reisine, T.D. (1984) *Science*, **224**, 452.

118. Aguilera, G., Harwood, J.P., Wilson, J.X., Morell, J., Brown, J.H. and Catt, K.J. (1983) *J. Biol. Chem.*, **258**, 8039.

119. Billard, W., Ruperto, V., Crosby, G., Iorio, L.C. and Barnett, A. (1984) *Life Sci.*, **35**, 1885.

120. Sidhu, A., van Olne, J.C., Dandridge, P., Kaiser, C. and Kebabian, J.W. (1986) *Eur. J. Pharmacol.*, **128**, 213.

121. Kebabian, J.W., Agui, T., van Oene, J.C., Shigematsu, K. and Soavedra, J.M. (1986) *Trends in Pharm. Sci.*, **7**, 96.

122. Sunahara, R.K., Niznik, H.B., Weiner, D.M., Stormann, T.M., Braun, M.R., Kennedy, J.L., Gelernter, J.E., Rozmahel, R., Yang, Y., Israel, Y., Seeman, P. and O'Dowd, B.F. (1990) *Nature*, **347**, 80.

123. Deary, A., Gingrich, J.A., Falardeau, P., Fremeau, R.T., Jr., Bates, M.D. and Carran, M.G. (1990) *Nature*, **347**, 72; also Zhou, Q.-Y. *et al.* (1990) *Nature*, **347**, 76.

124. Dahl, K.D. and Hsu, A.J.W. (1988) in *Hormones and their Actions Part II* (B.A. Cooke, R.J.B. King and H.F. van der Molen eds). Elsevier, Amsterdam, p. 181.

125. Tsutsui, K. (1991) *Endocrinology*, **128**, 477.

126. Exton, J.H. (1988) in *Hormones and their Actions Part II* (B.A. Cooke, R.J.B. King and H.J. van der Molen eds). Elsevier, Amsterdam, p. 231.

127. Wright, D.E., Horuk, R. and Rodbell, M. (1984) *Eur. J. Biochem.*, **141**, 63.

128. Ejalbert, A., Rasolonjanahany, R., Moyse, E., Kordon, C. and Epelbaum, J. (1983) *Endocrinology*, **113**, 822.

129. Czernick, A.J. and Petrack, B. (1983) *J. Biol. Chem.*, **258**, 5525.

130. Koch, B. and Schonbruns, A. (1984) *Endocrinology*, **114**, 1784.

131. Gajtkowski, G.A., Norris, D.B., Rising, T.J. and Wood, T.P. (1983) *Nature*, **304**, 65.

132. Mate, L., MacLellan, D.G. and Thompson, J.C. (1987) in *Gastrointestinal Endocrinology* (J.C. Thompson, G.H. Greeley, P.L. Rayford and C.M. Townsend eds). McGraw Hill, New York, p. 361.

133. Grantz, I., Schaffer, M., DelValle, J., Logsdon, C., Campbell, V., Uhler, M. and Yanada, T. (1991) *Proc. Natl. Acad. Sci. USA*, **88**, 429.

134. Birdsall, N. (1991) *Trends in Pharm. Sci.*, **12**, 9.

135. Cooke, B.A. and Rommerts, F.F.G. (1988) in *Hormones and their Actions Part II* (B.A. Cooke, R.J.B. King and H.J. van der Molen eds). Elsevier, Amsterdam, p. 155.

136. Jard, S. (1980) in *Cellular Receptors* (D. Schulster and A. Levitski eds). John Wiley and Sons, Chichester, p. 253.

137. Sheppard, G.M. (1988) *Neurobiology*, Oxford University Press, Oxford.

138. Karapalis, A.C. and Powell, W.S. (1981) *J. Biol. Chem.*, **256**, 2414.

139. Opmeer, F.A., Adolfs, M.J.P. and Bonta, I.L. (1983) *Biochem. Biophys. Res. Commun.*, **114**, 155.

140. Coleman, R.A., Humphrey, P.P.A., Kennedy, I. and Lumley, P. (1984) *Trends Pharmacol. Sci.*, **5**, 303.

141. Mate, L., Beachamp, R.D. and Thompson, J.C. (1987) in *Gastrointestinal Endocrinology* (J.C. Thompson, G.H. Greeley, P.L. Rayford and C.M. Townsend eds). McGraw Hill, New York, p. 372.

142. McCracken, J.A. and Schramm, W. (1988) in *Biology and Chemistry of Prostaglandins and Related Eicosanoids* (P.B. Curtis-Prior ed.). Churchill-Livingstone, New York, p. 425.

143. Laburthe, M., Breant, B. and Rouyer-Ferrard, C. (1984) *Eur. J. Biochem.*, **139**, 181.

144. Guild, S. and Drummond, A.H. (1984) *Biochem. J.*, **221**, 789.

145. Khalil, T., Alinder, G. and Rayford, P.L. (1987) in *Gastrointestinal Endocrinology* (J.C. Thompson, G.H. Greeley, P.L. Rayford and C.M. Townsend eds). McGraw Hill, New York, p. 260.

146. Evans, T., Smith, M.M., Tanner, L.I. and Harden, T.K. (1984) *Mol. Pharmacol.*, **26**, 395.

147. Haga, T. and Haga, K. (1990) in *Receptor-Effector Coupling: A Practical Approach* (E.C. Hulme ed.). IRL Press, Oxford University Press, Oxford, p. 83.

148. Exton, J.H. (1982) *Trends Pharm. Sci.*, **3**, 111.

149. Pimoule, C., Scatton, B. and Langer, S.Z. (1983) *Eur. J. Pharmacol.*, **95**, 79.

150. Bylund, D.B. (1988) *Trends Pharm. Sci.*, **9**, 356.

151. Harrison, K.J., Pearson, W.R. and Lynch, K.R. (1991) *Trends Pharm. Sci.*, **12**, 62.

152. Lamasney, J.W., Lorenz, W., Allen, L.F., King, K., Regan, J.W., Young-Feng, T.L., Caron, M.G. and Lefkowitz, R.J. (1990) *Proc. Natl. Acad. Sci.*, **87**, 5094.

153. Senagles, S., Benovic, J.L., Amlaiky, N., Unson, C., Milligan, G., Vinitsky, R., Spiegel, A.M. and Caron, M.G. (1987) *J. Biol. Chem.*, **262**, 4860.

154. Strange, P.G. and Williamson, R.A. (1990) in *Receptor Biochemistry: A Practical Approach* (E.C. Hulme ed.). IRL Press, Oxford University Press, Oxford, p. 79.

155. Mosberg, H.I., Hurst, R., Hruby, V.J., Gee, K., Yamamura, H.I., Galligan, J.J. and Burks, T.F. (1983) *Proc. Natl. Acad. Sci. USA* , **80**, 5871.

156. Miller, R.J. (1984) *Trends Neurochem. Sci.*, **7**, 184.

157. Demoliou-Mason, C.D. and Barnard, E.A. (1990) in *Receptor Biochemistry: A Practical Approach* (E.C. Hulme ed.). IRL Press, Oxford University Press, Oxford, p. 99.

158. Vanderwel, M., Lum, D.S. and Haslam, R.J. (1983) *FEBS Lett.*, **164**, 340.

159. Chardonnens, D., Lang, V., Capponi, A.M. and Vallotton, M.B. (1989) *J. Cardiovasc. Pharmacol.*, **14**, (Suppl. 6), S39.

160. Vallotton, M.B. (1991) *Mol. Cell. Endocr.*, **78**, C73.

161. Ullrich, A., Coussens, L., Hayflick, J.S., Dull, T.J., Gray, A., Tam, A.W., Lee, J., Yarden, Y., Libermann, T.A., Schlessinger, J., Downward, J., Mayes, E.L.V., Whittle, N., Waterfield, M.D. and Seeburg, P.H. (1984) *Nature*, **309**, 418.

162. Downward, J., Yarden, Y., Mayes, E., Scrace, G., Totty, N., Stockwell, P., Ullrich, A., Schlessinger, J. and Waterfield, M.D. (1984) *Nature*, **307**, 521.

163. Ullrich, A., Bell, J.R., Chen, E.Y., Herrera, R., Petruzelli, L.M., Dull, T.J., Gray, A., Coussens, L., Liao, Y.-C., Tsubokawa, M., Mason, A., Seeburg, P.H., Grunfeld, C., Rosen, O.M. and Ramachandran, J. (1985) *Nature*, **313**, 756.

164. Ebina, Y., Ellis, L., Jarnagin, K., Edery, M., Graf, L., Clauser, E., Ou, J.-H., Masiarz, F., Kan, Y.W., Goldfine, I.D., Roth, R.A. and Rutter, W.J. (1985) *Cell*, **40**, 747.

165. Bonner, T.I., Buckley, N.J., Young, A.C. and Brann, M.R. (1987) *Science*, **237**, 527.

166. Regan, J.W., Kobilka, T.S., Yang-Feng, T.L., Caron, M.G., Lefkowitz, R.J. and Kobilka, B.K. (1988) *Proc. Natl. Acad. Sci. USA* , **85**, 6301.

167. Bunzow, J.R., Van Tol, H.M.H., Grandy, D.K., Albert, P., Salmon, J., Christie, M., Machida, C.A., Neve, K.A. and Civelli, O. (1988) *Nature*, **336**, 783.

168. Fargin, A., Raymond, J.R., Lohse, M.J., Kobilka, B.K., Caron, M.G. and Lefkowitz, R.J. (1988) *Nature*, **335**, 358.

169. Masu, Y., Nakayama, K., Tamaki, H., Harada, Y., Kuno, M. and Nakanishi, S. (1987) *Nature*, **329**, 836.

170. Julius, D., Dermott, A.B., Axel, R. and Jessell, T.M. (1988) *Science*, **241**, 558.

171. Lubbert, H., Hoffman, B., Snutch, T.P., van Dyke, T., Levine, A.J., Hartig, P.R., Lester, H.A. and Davidson, N. (1987) *Proc. Natl. Acad. Sci. USA* , **84**, 4332.

172. Zhou, Q.-Y., Grandy, D.K., Thambi, L., Kushner, J.A., van Tol, H.H.M., Cone, R., Pribnow, D., Salon, J., Bunzow, J.R. and Civelli, O. (1990) *Nature*, **347**, 76.

173. Schulster, D. (1988) in *Radiochemicals in Biomedical Research* (E.A. Evans and K.G. Oldham eds) Critical Reports Applied Chemistry, Vol. 24. Wiley and Sons, Chichester, p. 94.

174. Rojas, F.J. and Birnbaumer, L. (1985) *J. Biol. Chem.*, **260**, 7829.

175. Luborsky, J.L., Dorflinger, L.J., Wright, K. and Behrman, H.R. (1984) *Endocrinology*, **115**, 2210.

176. Kikuchi, M. and Ishii, S. (1989) *Biol. Reprod.*, **41**, 1047.

177. Zhang, Q.Y. and Menon, K.M. (1989) *Proc. Natl. Acad. Sci. USA*, **86**, 8294.

178. May, J.V., Buck, P.A. and Schomberg, D.W. (1987) *Endocrinology*, **120**, 2413.

179. Lin, M.T. and Rao, C.V. (1978) *Mol. Cell. Endocr.*, **9**, 311.

180. Tomoi, M., Matsuo, M., Ono, T. and Shibayama, F. (1990) *Prostaglandins*, **39**, 113.

181. McIlhinney, R.A.J. and Schulster, D. (1975) *J. Endocrinol.*, **64**, 175.

182. Yanagibashi, K., Kamiya, N., Ling, G. and Matsuba, M. (1978) *Endocrinol. (Jpn)*, **25**, 545.

183. Buckley, D.I. and Ramachandran, J. (1981) *Proc. Natl. Acad. Sci. USA*, **78**, 7431; and Ramachandran, J. (1984–5) *Endocrin. Revs*, **10** (3, 4), 347.

184. Bruni, G., Dal-Pra, P. and Segre, G. (1989) *Drugs Exp. Clin. Res.*, **15**, 155.

185. Gallo-Payet, N. and Escher, E. (1985) *Endocrinology*, **117**, 38.

186. Bost, K.L. and Blalock, J.E. (1986) *Mol. Cell. Endocr.*, **44**, 1.

187. Brown, E.M., Aurbach, G.D., Hauser, D. and Troxler, F. (1976) *J. Biol. Chem.*, **251**, 1232.

188. Leurs, R., Beusenberg, F.D., Bast, A., van-Amsterdam, J.G. and Timmerman, H. (1990) *Inflammation*, **14**, 421.

189. Tiong, A.H. and Richardson, J.S. (1989–90) *J. Recept. Res.*, **9**, 495.

190. Harvey, S., Attardo, D. and Baidwan, J.S. (1990) *J. Mol. Endocrinol.*, **4**, 213.

191. Bilezikjian, L.M., Seifert, H. and Vale, W. (1986) *Endocrinology*, **118**, 2045.

192. Ikuyama, S., Natori, S., Nawata, H., Kato, K., Ibayashi, H., Kariya, T., Sakai, T., Rivier, J. and Vale, W. (1988) *J. Clin. Endocrinol. Metab.*, **66**, 1265.

193. Glossman, H., Baukal, A., Aguilera, G. and Catt, K.J. (1985) *Methods Enzymol.*, **109**, 110; and Viard, I., Rainey, W.E., Capponi, A.M., Begeot, M. and Saeż, J.M. (1990) *Endocrinology*, **127**, 2071.

194. Lewis, N.P. and Ferguson, D.R. (1989) *J. Endocrinol.*, **122**, 499.

195. Sutter, A., Riopelle, R.J., Harris-Warwick, R.M. and Shooter, E.M. (1979) *J. Biol. Chem.*, **254**, 5972.

196. Olender, E.J., Wagner, B.J. and Stach, R.W. (1981) *J. Neurochem.*, **37**, 436.

197. Vale, R.D. and Shooter, E.M. (1985) *Methods Enzymol.*, **109**, 21.

198. Schechter, A.L. and Bothwell, M.A. (1981) *Cell*, **24**, 867.

199. Koo, P.H. and Stach, R.W. (1989) *J. Neurosci. Res.*, **22**, 247.

200. Mayo, K.H., Nunez, M., Burke, C., Starbuck, C., Lauffenburger, D. and Savage, C.R. (1989) *J. Biol. Chem.*, **264**, 17838.

201. Reynolds, R.K., Tatavera, F., Roberts, J.A., Hopkins, M.P. and Menon, K.M. (1990) *Cancer*, **66**, 1967.

202. Kano-Sueoka, T., King, D.M., Fisk, H.A. and Klug, S.J. (1990) *J. Cell Physiol.*, **145**, 543.

203. Bowen-Pope, D.F. and Ross, R. (1982) *J. Biol. Chem.*, **257**, 5161.

204. Duan, D.S., Pazin, M.J., Fretto, L.J. and Williams, L.T. (1991) *J. Biol. Chem.*, **266**, 413.

205. Plouet, J. and Moukadiri, H. (1990) *J. Biol. Chem.*, **265**, 22071.

206. Talavera, F., Reynolds, R.K., Roberts, J.A. and Menon, K.M. (1990) *Cancer Res.*, **50**, 3019.

207. Penhoat, A., Chatelain, P.G., Jaillard, C. and Saez, J.M. (1988) *Endocrinology*, **122**, 2518.

208. Adashi, E.Y., Resnick, C.E., Hernandez, E.R., Svoboda, M.E. and van Wyk, J.J. (1988) *Endocrinology*, **122**, 1383.

209. Valent, P., Besemer, J., Muhm, M., Majdic, O., Lechner, K. and Bettelheim, P. (1989) *Proc. Natl. Acad. Sci. USA*, **86**, 5542.

210. Lopez, A.F., Eglinton, J.M., Gillis, D., Park, L.S., Clark, S. and Vadas, M.A. (1989) *Proc. Natl. Acad. Sci. USA*, **86**, 7022.

211. Nilsson, A., Lindahl, A., Eden, S. and Isaksson, O.G. (1989) *J. Endocrinol.*, **122**, 69.

212. Litwack, G. ed. (1990) *Receptor Purification*, Vol. 1, *Receptors for CNS Agents, Pituitary Growth Factors, Hormones and Related Substances*. Humana Press, New Jersey.

213. Limbird, L.E. ed. (1988) *The α_2-Adrenergic Receptors*. Humana Press, New Jersey.

214. Perkins, J.D. ed. (1991). *The β-Adrenergic Receptors*. Humana Press, New Jersey.

215. Brown, J.H. ed. (1989) *The Muscarinic Receptors*. Humana Press, New Jersey.

216. Sanders-Bush, E. ed. (1988) *The Serotonin Receptors*. Humana Press, New Jersey.

217. Pasternak, G.W. ed. (1988) *The Opiate Receptors*. Humana Press, New Jersey.

218. Sporn, N.B. and Roberts, A.B. eds (1991) *Peptide Growth Factors and their Receptors*. Springer Verlag, New York.

219. Cunningham, B.C., Henner, D.J. and Wells, J.A. (1990) *Science*, **247**, 1461.

CHAPTER 9

INHIBITORS

D. Rickwood

1. INTRODUCTION

Inhibitors have been one of the key tools in determining pathways at both the molecular and cellular levels. Inhibitors can be divided into two broad groups, those that are very specific (e.g. leupeptin) which are very specific by virtue of a very defined interaction with the active site of an enzyme and those which inhibit classes of proteins (e.g. diisopropylphosphofluoridate inhibits a wide range of proteases with serine at the active site). When using the latter class of inhibitors it is important to be certain that the effect that you are observing is the result of the inhibition of the enzyme or system of interest and not because of some other, secondary effect. In addition, often an inhibitor may have a different effect *in vivo* as compared with *in vitro*. An example of this is α-amanitin which is a specific inhibitor of RNA polymerase II that synthesises mRNA. *In vitro* this inhibitor has no effect on RNA polymerase I but *in vivo* it inhibits the synthesis of rRNA as a secondary effect of the inhibition of RNA polymerase II. The other feature is that the specificity of inhibitors can be very variable, this is especially true of comparisons between higher and lower eucaryotes. Hence, if you wish to use a particular type of inhibitor in a system that has not been well characterized then it would be wise to do so, as otherwise there may be doubts about the validity of the data obtained.

2. ABBREVIATIONS

Abbreviations: Anhyd., anhydrous; ATPase, adenine triphosphatase; cAMP, cyclic adenosine 3′,5′-monophosphate; $CHCl_3$, trichloromethane; CNS, central nervous system; concs, concentrations; dCTP, deoxycytidine triphosphate; DMF, dimethylformamide; DMSO, dimethylsulfoxide; DNA, deoxyribonucleic acid; EtOH, ethanol; HBr, hydrogen bromide; HnRNA, heterogenous ribonucleic acid; insol., insoluble; HMG-CoA, hydroxy methylglutanate-CoA; Me ester, methyl ester; MeOH, methanol; Mol. wt, molecular weight; mRNA, messenger ribonucleic acid; NAD, nicotinamide adenine dinucleotide; NADH, nicotinamide adenine dinucleotide (reduced form); RNA, ribonucleic acid; rRNA, ribosomal ribonucleic acid; sl. sol., slightly soluble; sol., soluble; sp. sol., sparingly soluble; tRNA, transport ribonucleic acid; UTP, uridine triphosphate; v. sol., very soluble; v. sl. sol., very slightly soluble; −ve, negative; +ve, positive; ♀, female; ♂, male.

Molecular weights have been given to the nearest whole number.

Table 1. Inhibitors of nucleic acids and protein synthesis

Name (and synonym)	Mol. wt	Solubility	Site and nature of action	Reference
Compounds affecting nucleic acids				
Actinomycin D	1255	V. sol. in acetone, $CHCl_3$; sol. in MeOH, EtOH, mineral acids; sl. sol. in water	Inhibits DNA-primed RNA polymerase by complexing with DNA, via deoxyguanosine residues; at higher concs DNA polymerase is inhibited; active in mammalian systems: potent antitumor agent	1
Adriamycin	543	Sol. in water, MeOH	Inhibits DNA and RNA synthesis to about the same extent; cytotoxic antitumor agent; damages DNA	1
α-Amanitin	903	Sol. in water, MeOH, EtOH	Inhibits eucaryote RNA polymerase II at 10^{-8}–10^{-9} M, polymerase III at 10^{-4}–10^{-5} M; does not inhibit either polymerase I or bacterial polymerase	2
Amethopterin (Methotrexate; 4-amino-N^{10}-methylpteroyl-glutamic acid)	454	Sol. in dilute HCl	Folic acid antagonist; reversed by folic acid, and by 5-formyltetrahydrofolic acid; inhibits dihydro-folate reductase inhibiting purine synthesis	3
Arabinosyl-cytosine (1-β-D-arabino-furanosyl-cytosine; cytabarine; cytosine-β-D-arabinofuranoside; cytosine arabinoside; AraC)	243	Sol. in water	Selectively inhibits DNA synthesis in mammalian cells (L-cells, 97% at 4×10^{-7} M without effect on RNA synthesis); bacteria are generally insensitive; triphosphate may act as competitive dCTP analogue and incorporation may cause slow chain extension	4

Table 1. Continued

Name (and synonym)	Mol. wt	Solubility	Site and nature of action	Reference
Azaserine (*O*-Diazoacetyl-L-serine)	173	V. sol. in water; aqueous solutions are most stable at pH 8, can be heated at 100°C in neutral aqueous solution for 5 min	Active against several *Clostridium* spp., *Mycobacterium tuberculosis* and rickettsiae; prevents transfer of amide group from glutamine to formylglycinamide ribotide in purine synthesis (inactivates phosphoribosyl-formylglycinamidine synthetase irreversibly by covalent attachment to –SH group); possesses antitumor activity; a few fungi, including several yeasts, are inhibited by azaserine	1
Bleomycin	A_2: 1415 B_2: 1425	V. sol. in water	Inhibits DNA synthesis > RNA; causes single-strand scission of DNA *in vivo* and *in vitro*, enhanced by reducing agents and Fe^{2+}; requires O_2 and reducing agent	5
Cordycepin (3′-Deoxyadenosine)	252 Monohydrate: 262.2	Sol. in water	Selectively inhibits polyadenylation of HnRNA in HeLa cells; inhibits 45S rRNA synthesis, antitumor, cytostatic agent in eucaryotic cells; reversed by adenosine, but not by 2′-deoxyadenosine	4
Daunomycin (Daunorubicin)	527 Hydrochloride: 563	HCl: sol. in water; MeOH; insol. in $CHCl_3$	Action as for adriamycin	1
Distamycin A (Stallimycin)	481 HCl: 518.1	Sol. in water	Inhibits DNA and RNA synthesis; active against DNA viruses; antitumor activity; binds to A:T regions of helical DNA; inhibits various DNA and RNA polymerases *in vitro*	6

Table 1. Continued

Name (and synonym)	Mol. wt	Solubility	Site and nature of action	Reference
Ethidium bromide (2,7-Diamino-10-ethyl-9-phenyl-phenanthridinium bromide; homidium bromide)	394	Sol. in water, EtOH, MeOH	Potent inhibitor of DNA synthesis *in vivo*; inhibits DNA synthesis by cell-free DNA polymerase and RNA synthesis by cell-free DNA-dependent RNA polymerase; it intercalates with DNA; reverses supercoiling of circular DNA	1
5-Fluorouracil (5-Fluoro-2,4-dihydroxypyrimidine)	130	Sol. in water	Metabolized to fluorodeoxyuridylic acid which inhibits thymidylic acid synthetase and to fluorouridine triphosphate which is incorporated into RNA in the place of UTP; it inhibits DNA synthesis; causes translation errors from mRNA to protein	1
Hydroxyurea (*N*-Hydroxyurea)	76	Sol. in water	Specific inhibitor of DNA synthesis *in vivo*; RNA and protein synthesis not affected; mechanism is via inhibition of deoxyribonucleotide synthesis from ribonucleotides; antiviral and antineoplastic; causes chromosome breakage; DNA repair synthesis not inhibited; actions reversible in eucaryotes, irreversible in bacteria	1
Mitomycin C	334	Sol. in water; readily decomposed by light	Inhibits DNA synthesis by cross-linking DNA strands; antitumour activity and strong bactericidal action against both Gram-positive and Gram-negative organisms and acid-fast bacilli	1
Netrospin (Congocidine)	430 Disulfate: 626	Disulfate: sl. sol. in water (0.5 mg ml^{-1} at 25°C); insol. in organic solvents	Inhibits DNA and RNA synthesis; binds to helical DNA at A–T rich regions (not by intercalation)	6

Table 1. Continued

Name (and synonym)	Mol. wt	Solubility	Site and nature of action	Reference
Phenethyl alcohol (2-Phenylethanol)	122	Sol. in EtOH; 1.6^{20} water	Inhibits DNA synthesis without affecting protein and RNA synthesis at 0.25%; but at higher concs protein and RNA synthesis inhibited and cells no longer elongate; is bacteriostatic to a wide variety of bacterial species	1
Rifampicin (Rifampin)	823	V. sl. sol. in water	Inhibits transcription in bacteria at extremely low concs; binds to β-subunit of bacterial RNA polymerase and inhibits chain initiation	1
Vinblastin sulfate	909	Sol. in water	Inhibits karyokinesis and secretion; interferes with microtubule organization; inhibits RNA synthesis; effects not easily reversible	1
Compounds affecting protein synthesis				
Aurintricarboxylic acid (ATA)	422	Sol. in water	In low concs, reported to be a relatively specific inhibitor of initiation of protein synthesis; at high concs, elongation is affected	7
Chloramphenicol (CAP; chloromycetin)	323	0.25 water; v. sol. in MeOH, EtOH	Inhibits bacterial protein synthesis by blocking peptidyl transferase reaction; inhibits mitochondrial and chloroplast protein synthesis; irreversible toxicity to animal cells and humans at very high concns may be due to inhibition of DNA synthesis	8
Cycloheximide (Actidione)	281	Sol. in water, EtOH, MeOH; unstable in alkali	Inhibits cytoplasmic eucaryotic protein synthesis but not procaryotic; proposed that it inhibits initiation, and elongation	9

Table 1. Continued

Name (and synonym)	Mol. wt	Solubility	Site and nature of action	Reference
Emetine (6′,7′,10,11-Tetramethoxyemetan)	481 Dihydrochloride: 554	V. sl. sol. in water; sol. in EtOH, acetone; dihydrochloride: v. sol. in water	Selectively inhibits protein synthesis in eucaryotic cells and extracts; acts on 40S ribosomal subunit site and inhibits elongation; DNA synthesis affected	10
Erythromycin (Ery)	734 (Ery A)	V. sol. in MeOH < EtOH; slowly sol. in water (2 mg ml^{-1})	Inhibitor of procaryotic protein synthesis by inhibiting transpeptidation so stopping elongation	1
Puromycin (Stylomycin)	471	Free base, sp. sol. in water and organic solvents; dihydrochloride and monosulfate, readily sol. in water	Inhibits protein synthesis; prevents growth of bacteria, protozoa, algae and mammalian cells; causes premature release of partially formed protein chains from ribosomes by acting as an analog of acceptor aminoacyl-tRNA	1
Streptomycin	582	Sol. in water	Inhibits initiation, elongation and termination of protein synthesis in procaryotes and induces misreading; binds to small ribosomal subunit	1
Chlorotetracycline (Aureomycin)	479 Hydrochloride: 515	Sl. sol. in water (0.5 mg ml^{-1}); v. sol. > pH 8.5 Hydrochloride: sol. in water (8.6 mg ml^{-1}), MeOH	Inhibitor of protein synthesis in procaryotes and eucaryotes; preferentially accumulated by bacteria, conferring therapeutic value; inhibits binding of aminoacyl-tRNA to ribosomes; so inhibiting elongation	11
Oxytetracycline (Terramycin)	460 Dihydrate: 496 HCl: 497	Sol. in water, EtOH Hydrochloride: v. sol. in water (1 mg ml^{-1}); aqueous solutions of hydrochloride stable 4 weeks, pH 3–9, at 5°C		

Table 2. Inhibitors of membrane function

Name (and synonym)	Mol. wt	Solubility	Site and nature of action	Reference
Amphotericin B (Polyene antibiotic; amphoteric polyene)	924 Me ester hydrochloride: 975	V. sol. in water at pH 7; sl. sol. in DMF (2–4 mg ml^{-1}), DMSO (30 mg ml^{-1}); sol. at pH 2 and 11 Me ester HCl: sol. in water (> 75 mg ml^{-1})	Induces loss of low mol. wt substances from cells with low selectivity; probably forms channels by complexing with membrane cholesterol; mitochondria not affected; lyses red blood cells; effective at low concs	12
Atractyloside (Potassium atractylate)	Atractyloside K_2 salt: 803 Carboxy-atractyloside: 771	Sol. in EtOH; insol. in water	Prevents movement of adenine nucleotides across inner mitochondrial membrane by displacing nucleotides from translocase protein in the membrane; carboxyatractyloside is a specific inhibitor of adenine nucleotide transport but non-competitive towards nucleotides	13
Bongkrekic acid	487	Sol. in dilute alkali	Prevents movement of adenine nucleotides across inner mitochondrial membrane by preventing release of nucleotides from carrier protein (translocase) on the matrix side	13, 14
α-Bungarotoxin	7983		Neurotoxin, acting postsynaptically via blockade of the acetylcholine receptor	15
β-Bungarotoxin	20500		Neurotoxin, modifies release of neurotransmitter from mammalian motor-nerve terminals; inhibits Ca^{2+} accumulation in subcellular fractions of brain	16

Table 2. Continued

Name (and synonym)	Mol. wt	Solubility	Site and nature of action	Reference
Cytochalasin	A: 478 B: 480	Insol. in water; sol. in acetone	Inhibits cell motility, phagocytosis and the division of cytoplasm; causes nuclear extrusion; modifies, reversibly, cell microfilaments; actin bundle patterns rapidly disorganized by Cytochalasin B, while microtubules are not affected; cytochalasin A inhibits tubulin and actin assembly *in vitro* and decreases the colchicine binding of tubulin, through an irreversible action on –SH groups of proteins	17, 18
Gramicidins A, B, C (Linear gramicidins)	1884		Channel-forming antibiotics, rendering membranes permeable to protons and alkali metal cations; hemolyse red blood cells; cause K^+–H^+ exchange in mitochondria	19
Gramicidin S	1141 Dihydro-chloride: 1214	Sol. in EtOH; insol. in water	Uncouples oxidative phosphorylation in mitochondria	1
Ouabain (G-Strophanthin)	585 Hydrate: 729	Sol. in water and EtOH	Inhibits cation transport; classical inhibitor of Na^+–K^+ ATPase (10^{-6} M)	1
Tetrodotoxin (TTX)	319	Sol. in water, EtOH and in dilute acetic acid; stable in solution when at pH 4–5 and 0°C; unstable in acid and alkaline solutions; destroyed by boiling at pH 2 and below	Neurotoxin inhibiting the action potential of nerve membranes at 10^{-9}–10^{-7} M, it blocks the early transient inward flow of Na^+ ions necessary for the initial depolarization of the nerve membrane	1

Table 3. Inhibitors of mitochondrial and chloroplast functions

Name (and synonym)	Mol. wt	Solubility	Site and nature of action	Reference
Antimycin	A_1: 549 A_3: 521	A_1: v. sol. in acetone and EtOH; almost insol. in water A_3: v. sol. in acetone; sol. in EtOH, MeOH; almost insol. in water	Inhibits mitochondrial electron transport specifically between cytochromes *b* and *c*	1
Aurovertin B	460	Sol. in EtOH; insol. in water	Binds to mitochondrial ATPase, preventing phosphoryl group transfer and thereby oxidative phosphorylation; effective on isolated enzyme	20
CCCP (Carbonylcyanide *m*-chlorophenyl hydrazone)	205		High concs strongly inhibitory to electron transport	1
CCP (Carbonylcyanide phenylhydrazone; phenylhydrazone malanonitrile)	170	Sol. in EtOH, MeOH; sp. sol. in water ($< 10^{-4}$ M)	Uncouples oxidative phosphorylation and photophosphorylation	1
3-(*p*-Chlorophenyl)-1,1-dimethylurea (CMU; Monuron)	199	Sol. in MeOH and in acetone	Inhibits O_2 evolution step of photosynthesis at 10^{-6}–10^{-7} M; herbicide	1
Chloropromazine (Largactil)	Hydrochloride: 355 Free base: 319	V. sol. in water; sol. in EtOH and MeOH	Uncouples and inhibits oxidative phosphorylation at 10^{-3}–10^{-4} M	21

Table 3. Continued

Name (and synonym)	Mol. wt	Solubility	Site and nature of action	Reference
2,4-Dinitrophenol	184	Sl. sol. in EtOH; 0.56^{18} water	Uncouples oxidative phosphorylation at 10^{-5}–10^{-4} M by mediating proton conductance across inner mitochondrial membrane; stimulates mitochondrial ATPase; does not uncouple substrate-level phosphorylation	1
FCCP (Carbonylcyanide *p*-trifluoromethoxy-phenylhydrazone; *p*-CF_3O-CCP)	254	Sol. in EtOH, MeOH; sp. sol. in water ($<10^{-4}$ M)	Increases electrical conductivity of model phospholipid membranes; inactivated *in vitro* by vicinal aminothiols and dithiols	1
Hydrogen cyanide (Hydrocyanic acid)	27	Free acid, infinite in water, EtOH	Inhibits by complexing with metals in metallo-enzymes, e.g. cytochrome oxidase, at 10^{-3}–10^{-5} M; or by forming cyanhydrins with carbonyl groups	1
Hydrogen cyanide, K salt (Potassium cyanide)	65	Salt, v. sol. in water; sol. in EtOH		
Oleic acid	282	Infinite in EtOH; insol. in water	Uncouples oxidative phosphorylation; inhibits 2,4-dinitrophenol-stimulated ATPase; action reversed by adding serum albumin	1
Oligomycin B	805	Sol. in EtOH; 0.002 water	Inhibits mitochondrial ATPase (F_1), preventing phosphoryl group transfer; inhibits only membrane-bound, not isolated, form; inhibits Na^+–K^+ ATPase	22
Piericidin A	401	Sol. in EtOH	Inhibits NAD-linked substrate oxidation by mitochondria at same site as rotenone	23
Rotenone	394	Sol. in EtOH and in acetone; almost insol. in water	Inhibits NAD-linked substrate oxidation by mitochondria at oxygen side of NADH dehydrogenase	23

Table 4. Selected inhibitors of enzymes

Name (and synonym)	Mol. wt	Solubility	Site and nature of action	Reference
Alloxan (Mesoxalylurea)	Anhyd.: 142 Hydrate: 160	Sol. in water, EtOH	Cytotoxic, diabetogenic; at 10^{-3} M inhibits skeletal muscle hexokinase; inhibition reversed by cysteine; inhibits succinic dehydrogenase	1
Amphetamine (Benzedrine; 1-phenyl-2-amino-propane, (+)-α-methyl-phenylethylamine; (+)-isomer: dexamphetamine)	135 Sulfate: 368	Sol. in EtOH, acids, $CHCl_3$; sl. sol. in water Sulfate, sol. in water; sl. sol. in EtOH	Strong inhibitor of amine oxidase; central stimulant, peripheral sympathomietic	1
Antipain ([(S)-1-carboxy-2-phenylethyl]-carbamoyl-L-arginyl-L-valylargininal)	605	Di-HCl: sol. in water, MeOH, DMSO; sl. sol. in EtOH; insol. in $CHCl_3$	Inhibits cathepsin, chymotrypsin, papain, pepsin, plasmin and trypsin	24
Caffeine (1,3,7-Trimethyl-xanthine; theine; methyltheo-bromine)	194 Hydrate: 212 HCl: 267 Sulfate: 292	Sol. in hot water, EtOH, acetone and $CHCl_3$; sl. sol. in cold water	Inhibits cAMP phosphodiesterase; stimulates CNS; vasoconstrictor	1
Cerulenin (2,3-Epoxy-4-oxo-7,10-dodecadienamide)	223	Sol. in EtOH, acetone; sl. sol. in water Stable in neutral and acidic solutions	Inhibits fatty acid synthetase, HMG-CoA synthetase and prevents sterol synthesis; antifugal, antibacterial; inhibition of peptidase reported	25

Table 4. Continued

Name (and synonym)	Mol. wt	Solubility	Site and nature of action	Reference
p-Chloromercuri-benzene-sulfonic acid (PCMBS)	393	V. sol. in water	Enzyme inhibitor at 10^{-5} M; has high affinity for –SH groups in proteins	26
Chymostatin (*N*-[{(S)-1-carboxy-2-phenyl-ethyl}carbamoyl]-α-[2-iminohexahydro-4(S)-pyrimidyl]-L-glycyl-L-leucylphenylalaninal)	583	Sol. in acetic acid, DMSO; sl. sol. in water, EtOH	Inhibits chymotrypsin strongly, papain and cathepsins weakly, inactive on trypsin	27
Diethylpyrocarbonate (DEP; diethyl dicarbonate; ethoxyformic acid anhydrate)	162	V. sol. in EtOH and in organic solvents; decomposed by heating in water	Destroys enzymes by ethoxyformylation of proteins; has been used to inhibit nucleases during RNA extraction but known to destroy biological activity of RNA	28
Diisopropylphospho-fluoridate (Di-isopropylfluoro-phosphonate; DFP)	184	Sol. in organic solvents; 1.54^{25} water; aqueous solutions unstable	Inhibits serine esterases by covalent attachment to active site serine; pseudocholinesterase inhibitor at 10^{-7}–10^{-9} M, chymotrypsin and trypsin at 10^{-3}–10^{-5} M	1
Elastinal (*N*-[(S)-1-carboxyisopentyl]-carbamoyl-α-[2-iminohexahydro-4(S)-pyrimidyl]-L-glycyl-L-glutaminyl-L-alaninal)	513	Sol. in water, DMSO; sl. sol. in EtOH, acetone Unstable in alkaline solution	Strong inhibitor of elastase	29

Table 4. Continued

Name (and synonym)	Mol. wt	Solubility	Site and nature of action	Reference
Ephedrine (adrenalin; α-(1-methylamino-ethyl)-benzyl alcohol; 1-1-phenyl-2-methyl-aminopropanol)	165 HCl: 202	Sol. in water, EtOH, $CHCl_3$, oils HCl and sulfate, sol. in water, ETOH Aqueous solution stable to light, air and heat	Sympathomimetic; competitive inhibitor of choline dehydrogenase	1
Iodoacetamide	185	Sol. in water	Enzyme inhibitor at 10^{-3} M; reacts with –SH groups	1
Iodoacetic acid	186	Free acid, sol. in water, EtOH	Enzyme inhibitor at 10^{-3} M; reacts with –SH groups	1
Iodoacetic acid, Na salt	208	Salt, sol. in water		
Leupeptin (*N*-Propionyl [or acetyl]-L-leucyl-L-leucyl-L-arginal)	427 (acetyl)	HCl: sol. in water, EtOH, DMSO; insol. in acetone	Strongly inhibits cathepsin B, papain, plasmin and trypsin; cathepsins A and D, chymotrypsin and pepsin are insensitive	30
Pepstatin A	686	Sol. in MeOH, EtOH, DMSO; sl. sol. in water	Inhibits acid proteases strongly; inhibits renin	31
Phenylmethyl-sulfonyl fluoride (PMSF; α-toluene-sulfonyl fluoride)	174	Sol. in isopropanol and EtOH; hydrolyzed by water	Specific serine protease inhibitor; sulfonylates protein exclusively at the active site	1
Pyrazole (1,2-Diazole)	68	Sol. in water and in EtOH	Inhibits alcohol dehydrogenase	32
Theophylline (3,7-Dihydro-1,3-dimethyl-1*H*-purine-2,6-dione)	180	Sol. in hot water, alkaline hydroxides, dilute acids; 0.9 water; 1.25 EtOH	Inhibits cAMP phosphodiesterase competitively	33

Table 4. Continued

Name (and synonym)	Mol. wt	Solubility	Site and nature of action	Reference
TLCK (Tosyllysine chloromethylketone)	333 Hydrochloride: 369	Sl. sol. in water	Inactivates trypsin but not chymotrypsin; reacts stoichiometrically with the active center	1
TPCK (Tosylphenyl-alanine chloromethyl-ketone; L-(1-tosylamido-2-phenyl)-ethyl chloro-methylketone)	352	Sol. in MeOH; hydrolyzed in water	Specific inhibitor of chymotrypsin but not trypsin; reacts with the active center	1

Table 5. Selection of miscellaneous inhibitors

Name (and synonym)	Mol. wt	Solubility	Site and nature of action	Reference
Dihydro-β-erythriodine	275 HBr, 356	Base, sol. in water and in EtOH	Neuromuscular and ganglionic blocking agent; inhibits cholinergic synaptic transmission to Renshaw cells	1
Decamethonium (Deca-methylene-bis[tri-methylammonium]; Bromide: Syncurine)	Bromide: 418	Sol. in water, EtOH; aqueous solution stable	Neuromuscular blocking agent	1
Hexamethonium (Hexa-methylene-bis[tri-methylammonium]; hexameton chloride)	Chloride: 273	Sol. in water, EtOH	Ganglionic blocking agent	1

Table 5. Continued

Name (and synonym)	Mol. wt	Solubility	Site and nature of action	Reference
Phenoxybenzamine (Dibenzyline; dibenyline; *N*-(2-chloroethyl)-*N*-(1-methyl-2 phenoxy-ethyl)-benzylamine)	309	Sol. in HCl, EtOH, $CHCl_3$; sp. sol. in water	Adrenergic blocking agent	1
Propranolol (Inderal; propanolol)	259 HCl: 296	Sol. in EtOH; more stable at pH 3 HCl: sol. in water	β-Adrenergic blocking agent; abolishes oubain-induced arrhythmias	1
Prostaglandin E_1 (11a-15(S)-Dihydroxy-9-oxo-13-*trans*-prostenoic acid; PGE_1)	354	V. sl. sol. in water; sol. in MeOH, EtOH, $CHCl_3$; maximum stability between pH 6 and 7	Complex effects on reproductive, respiratory and cardiovascular systems and mobilizes free fatty acids from adipose tissues; antagonist of hypotensive and diuretic action of angiotensin	34
Strychnine	334 HCl: 407 Sulfate: 857	Sol. in $CHCl_3$; sl. sol. in EtOH; sp. sol. in water Hydrochloride and sulfate, sol. in water; sl. sol. in $CHCl_3$, EtOH	Blocks postsynaptic inhibition; convulsant	1

3. REFERENCES

1. Elliott, D.C., Elliot, W.H. and Jones, K.M. eds (1986) *Data for Biochemical Research (3rd edn)*. Oxford University Press, Oxford.

2. Wieland, T. and Faulstich, H. (1978) *Crit. Rev. Biochem.*, **5**, 185.

3. Nicolini, C. (1976) *Biochim. Biophys. Acta*, **458**, 264.

4. Suhadolnick, R.J. (1979) *Prog. Nucleic Acid Res. Mol. Biol.*, **22**, 193.

5. Hochenheim, J.S. (1979) *Antibiotics*, **5**, 124.

6. Zimmer, C. (1975) *Prog. Nucleic Acid Res. Mol. Biol.*, **15**, 285.

7. Chatterjee, N.K., Dickerman, H.W. and Beach, T.A. (1977) *Arch. Biochem. Biophys.*, **183**, 228.

8. Monro, R.E. and Vasquez, D. (1967) *J. Mol. Biol.*, **28**, 161.

9. Oleinick, N.L. (1977) *Arch. Biochem. Biophys.*, **182**, 171.

10. Jimenez, A., Carrasco, L. and Vazquez, D. (1977) *Biochemistry*, **16**, 4727.

11. Gordon, J. (1969) *J. Biol. Chem.*, **244**, 5680.

12. Norman, A.W., Spielvogel, A.M. and Wong, R.G. (1976) *Adv. Lipid Res.*, **14**, 127.

13. Vignais, P.V. (1976) *Biochim. Biophys. Acta*, **456**, 1.

14. Lauquin, G.J.M., Duplaa, A., Klein, G., Rousseau, A. and Vignais, P.V. (1976) *Biochemistry*, **15**, 2323.

15. Tu, A.T. (1973) *Ann. Rev. Biochem.*, **42**, 235.

16. Wagner, G.M., Mart, P.E. and Kelly, R.B. (1974) *Biochem. Biophys. Res. Commun.*, **58**, 475.

17. Pollard, T. and Rosenbaum, J. eds (1976) *Cell Motility*. Cold Spring Harbor Laboratory, New York, p. 403.

18. Himes, R.H. and Houston, L.L. (1976) *J. Supramol. Struct.*, **5**, 81.

19. Ovchinnikov, Y.A. (1979) *Eur. J. Biochem.*, **94**, 321.

20. Penefsky, H.S. (1979) *Adv. Enzymol.*, **49**, 223.

21. Dawkins, M.J.R., Judah, J.D. and Rees, K.R. (1959) *Biochem. J.*, **73**, 16.

22. Lardy, H., Reed, P. and Lin, C.C. (1975) *Fed. Proc.*, **34**, 1707.

23. Singer, T.P. and Gutman, M. (1971) *Adv. Enzymol.*, **34**, 79.

24. Suda, H., Aoyagi, T., Hameda, M., Takeuchi, T. and Umezawa, H. (1972) *J. Antibiotics*, **25**, 263.

25. Omura, S. (1976) *Bacteriol. Rev.*, **40**, 681.

26. Madsen, N.B. (1963) *Metabolic Inhibitors*, **2**, 119.

27. Tatsuta, K., Mikami, M., Fujmoto, K. and Umezawa, H. (1973) *J. Antibiotics*, **26**, 625.

28. Ehrenberg, L., Federsak, I. and Solymosy, F. (1976) *Prog. Nucleic Acid Res. Mol. Biol.*, **16**, 189.

29. Okura, A., Marishima, H., Takita, T., Aoyagi, T., Takeuchi, T. and Umegzawa, H. (1973) *J. Antibiotics*, **28**, 337.

30. Shimizu, B., Saito, A., Akira, T. and Tokawa, K. (1972) *J. Antibiotics*, **25**, 515.

31. Aoyagi, T., Marishima, H., Nishizawa, R., Kunimoto, S., Takendi, T. and Umezawa, H. (1972) *J. Antibiotics*, **25**, 689.

32. Teschke, R., Matsuzaki, S., Ohnishi, K., Decarli, L.M. and Lieber, C.S. (1977) *Clin. Exper. Res.*, **1**, 7.

33. Butcher, R.W. and Sutherland, E.W. (1962) *J. Biol. Chem.*, **237**, 1244.

34. Lee, J.B. ed. (1982) *Prostaglandins*. Elsevier, New York.

CHAPTER 10

CYCLIC NUCLEOTIDES

R. P. Newton

1. CYCLIC NUCLEOTIDE STRUCTURE, OCCURRENCE AND FUNCTION

Cyclic nucleotides constitute a specific category of compounds whose major physiological significance is derived from their activity as regulators of cellular metabolism. Cyclic nucleotides contain a purine or pyrimidine base linked to a sugar moiety, either β-D-ribose or β-D-2′-deoxyribose, to which a phosphate group is bonded; unlike the non-cyclic mononucleotides, which have the phosphate linked to the sugar at one of the 5′-, 3′- or 2′-positions, cyclic nucleotides contain a phosphate group linked at two positions, either the 3′- and 5′-positions or the 2′- and 3′- positions (*Figure 1*). This phosphodiester link renders the cyclic nucleotides more stable to acid hydrolysis and less polar than non-cyclic nucleotides.

Figure 1. Cyclic nucleotide isomers.

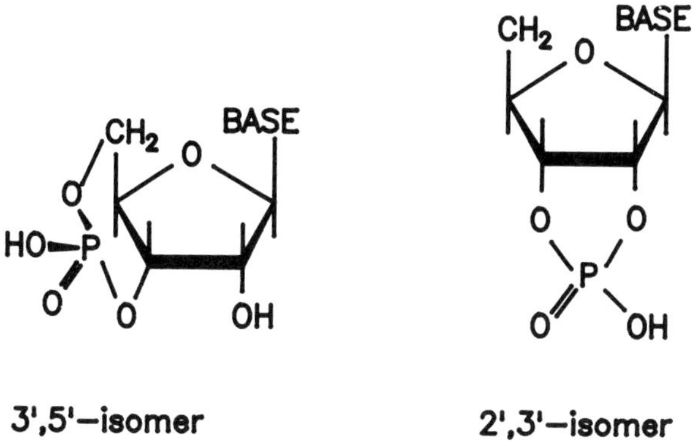

The 2′,3′-cyclic nucleotide isomers are merely products of nucleic acid breakdown and, as such, are not of great regulatory significance; the 3′,5′-cyclic nucleotides, with which this chapter is primarily concerned, are, on the other hand, key metabolic regulators, and the presence of the six-membered phosphate ring together with the five-membered ribose ring in these compounds produces a rigid bicyclic structure involving a half chair–chair conformation (1). At least eight 3′,5′-cyclic nucleotides are naturally occurring in mammalian and plant tissues (2–6), with cyclic AMP (cAMP), cyclic GMP (cGMP), cyclic CMP (cCMP) and cyclic UMP (cUMP) also being reported in species of bacteria (7) (*Figure 2*). cAMP was the first discovered (2) and has been by far the most extensively studied of these compounds, with the elucidation of the role of cAMP in mediating hormone action leading to the concept of biochemical second messengers (8). In the case of cAMP, the primary messenger is a hormone or neurotransmitter which binds to a specific receptor on the cell membrane surface, leading to a stimulation or inhibition of the enzyme adenylyl cyclase, which catalyzes

Figure 2. 3′,5′-cyclic nucleotides.

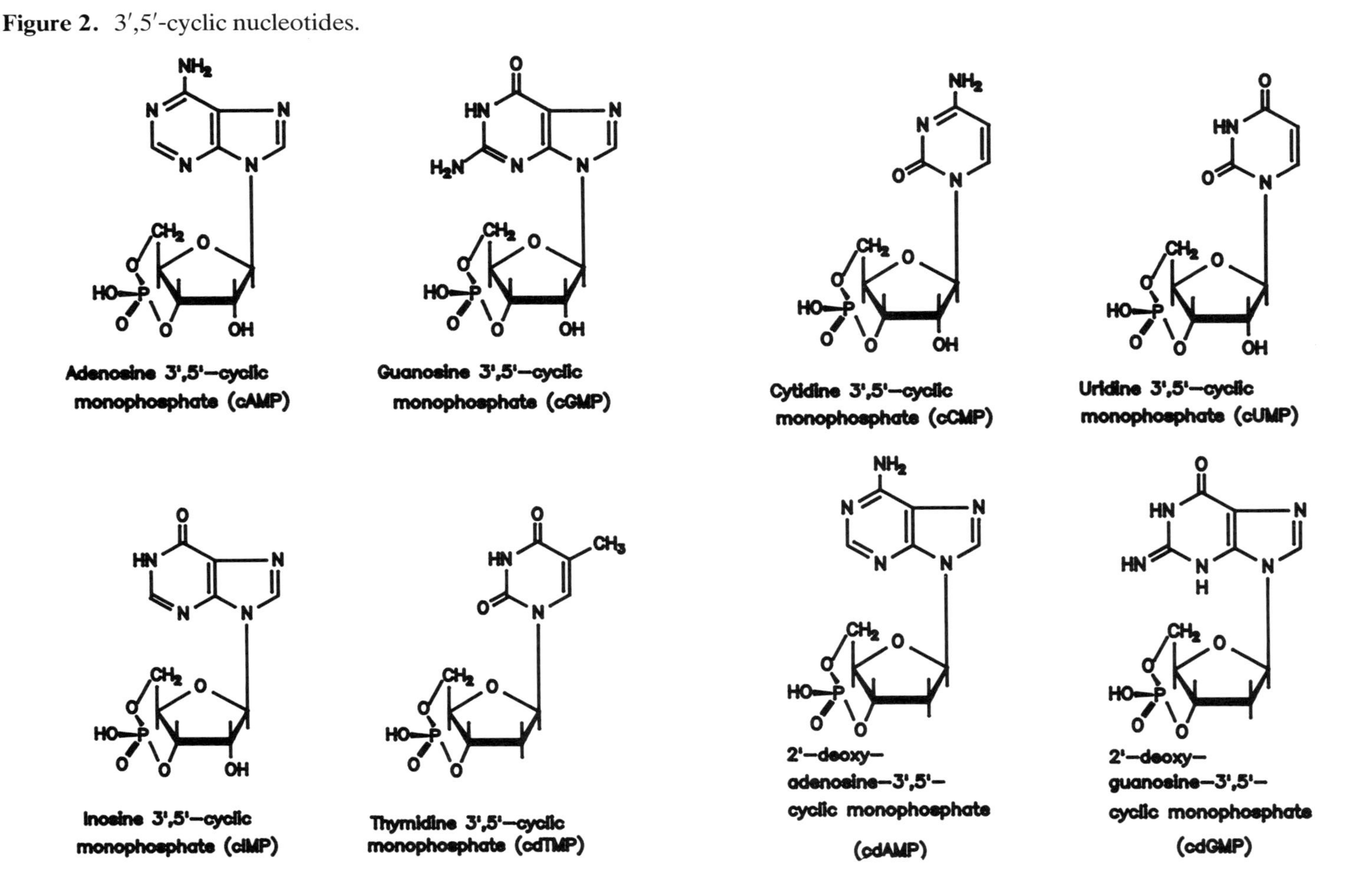

the conversion of adenosine 5′-triphosphate (ATP) to cAMP at a catalytic subunit sited on the inner side of the cell membrane. The cAMP so formed is released into the cell and elicits a response to the hormone by activating a cAMP-dependent protein kinase; this protein kinase exists as an inactive complex in the absence of cAMP, but on binding cAMP to regulatory subunits, the complex dissociates to release active catalytic subunits which, in turn, phosphorylate a variety of substrate proteins. This phosphorylation modifies the biological activity of these proteins and thereby produces the observed response to the primary messenger (*Figure 3*). Within the cell, cAMP is hydrolyzed by a cyclic nucleotide phosphodiesterase to the corresponding mononucleotide, AMP, thereby switching off the hormone response signal. Thus the level of intracellular cAMP is primarily the result of the balance between the adenylyl cyclase activity and the activity of a number of phosphodiesterase isoenzymes; in addition, GTP-binding proteins, commonly referred to as G-proteins, are involved in the transduction of the signal from the receptor unit across the cell membrane to the catalytic unit of adenylyl cyclase (9). Analogous cyclases and phosphodiesterases, capable of synthesis and hydrolysis of other naturally occurring cyclic nucleotides, also exist (5).

While there are a large number of metabolic processes regulated by cAMP (see *Table 9*) and a multiplicity of agonists stimulating the adenylyl cyclase system (8), cGMP appears to perform a much more restricted role, functioning in the transduction of the visual signal in the mammalian eye (10) and in mediating the action of atrial natriuretic peptide in the regulation of plasma water/electrolyte balance (11). No function for the other naturally occurring cyclic nucleotides has yet been established, although evidence exists that cCMP performs a role in the regulation of cell proliferation (12), while few reports relating to deoxy cAMP and deoxycyclic cGMP exist.

Figure 3. cAMP second messenger system.

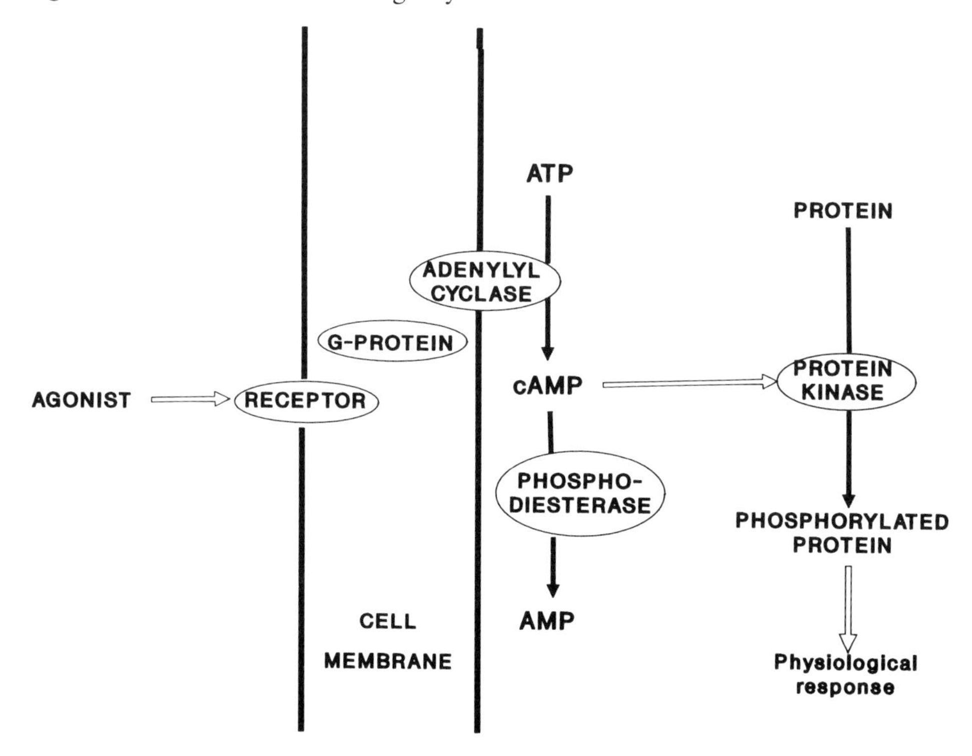

Cyclic Nucleotides

2. EXTRACTION OF CYCLIC NUCLEOTIDES

The major difficulty in determining the endogenous cyclic nucleotide concentrations of living tissue is ensuring that these levels are not altered during the extraction process. Most commonly used protocols therefore employ a rapid 'freeze-kill' procedure, utilizing silver or aluminum clamps prechilled in liquid N_2 (13, 14); the resultant material may be first freeze-dried, or directly homogenized in either ice-cold 10% aqueous trichloroacetic acid (9–12 volume:weight sample) or 0.6 M perchloric acid (6–9 volume:weight sample), dependent upon the method of assay selected (see Section 5), and the precipitated protein removed by centrifugation (14, 15). Perchloric acid extracts are neutralized with 1 M KOH and the resultant precipitate of $KClO_4$ removed by centrifugation at 4°C, or neutralized with octylamine and extracted with Freon (15); trichloroacetic acid preparations are extracted several times with two volumes of water-saturated diethyl ether to remove trichloroacetate. Neutralized extracts require storage at −20°C or lower to prevent changes in cyclic nucleotide content.

3. CHROMATOGRAPHIC SEPARATION OF CYCLIC NUCLEOTIDES

Prior to estimation of cyclic nucleotides in a tissue extract, chromatographic separation of the particular cyclic nucleotide to be assayed from other cyclic nucleotides, non-cyclic nucleotides, nucleosides and free bases may be necessary if the method of estimation is not sufficiently specific. Cation exchange, anion exchange and combined ion-exchange molecular sieve column chromatography can be utilized for preparative purposes, often with the inclusion of a [^{3}H]- or [^{14}C]-labeled cyclic nucleotide standard to estimate recoveries (*Table* 1). Column dimensions and elution volumes vary depending upon the size of the sample to be applied. Thin-layer chromatography provides a useful means of visual interpretation of cyclic nucleotide components within an analyte mixture, the most commonly used being polyethyleneimine (PEI) cellulose developed with LiCl (*Table 2*). While use of 50 mM LiCl as the developing solvent is adequate for most purposes, many variations are used, dependent upon the degree of resolution required. Three or four successive unidirectional developments with sodium formate, sodium acetate, KCl, LiCl, H_2O and acetic acid will yield complete resolution of the naturally occurring nucleotide/nucleoside components extracted as described in Section 2 (21).

Table 1. Column chromatography of cyclic nucleotides

Matrix	Form	Eluent	Ref.
Dowex 50	H^+	0.1 M HCl	16
Dowex 1	$HCOO^-$	0.5–4 M HCOOH	17
QAE Sephadex	$HCOO^-$	0.1 M $HCOONH_4$; 10^{-2} M HCl	16, 18
SP Sephadex	$HCOO^-$	10^{-2} M CH_3COOH	18
PEI cellulose	—	0.2–1 M LiCl	16
Aluminum oxide	—	0.1 M imidazole–50 mM HCl, pH 7	19
Boronate	—	0.5 M Hepes–NaCl, pH 8.45	20

Systems listed are designed to elute cyclic nucleotides rapidly and retain non-cyclic nucleotides, with the exception of SP Sephadex, which retains free bases and nucleotides, and boronate, which retains compounds possessing a *cis* diol group, i.e. free 2′- and 3′-OH groups on the ribose ring.

Table 2. Thin-layer chromatography of cyclic nucleotides; Rf values of cyclic nucleotides on PEI cellulose with 25, 50 and 75 mM LiCl as developing solvent

Cyclic nucleotide	LiCl		
	25 mM	50 mM	75 mM
cAMP	0.16	0.31	0.46
cGMP	0.13	0.27	0.44
cIMP	0.16	0.34	0.49
cCMP	0.38	0.58	0.72
cUMP	0.45	0.65	0.80
cdTMP	0.02	0.10	0.17

Compounds visualized as spots absorbing u.v. light when irradiated with light at 254 or 280 nm; the effect can be highlighted by spraying with 0.1% ethanolic 2′,7′-dichlorofluorescein and then drying the chromatogram. (Unpublished observations from the author's laboratory.)

HPLC provides a useful analytical tool for cyclic nucleotide studies, but preparative HPLC prior to cyclic nucleotide estimation can create difficulties of the eluting solvent interfering in the subsequent assay. Analytical HPLC can be utilized in ion-exchange, reverse-phase and paired-ion modes, and with purified samples the u.v. absorbance (see Section 4 and *Table 4*) can provide a direct means of estimation (*Table 3*).

4. U.V. ABSORPTION, MASS AND NMR SPECTRA CHARACTERISTICS OF CYCLIC NUCLEOTIDES

Methods used to estimate cyclic nucleotides (Section 5) do not have absolute specificity; thus when the cyclic nucleotide content of a particular tissue is being determined for the first time it is necessary to confirm the identity of the putative cyclic nucleotide by means of its charac-

Table 3. High-performance liquid chromatography of cyclic nucleotides

System	Retention times (min)					
	cAMP	cGMP	cIMP	cCMP	cUMP	cdTMP
Anion exchange	19	38	27	8	12	—
Reverse phase	17	11	—	6	9	—
Paired ion	21	15	18	9	11	13

Data for anion exchange obtained with Aminex A-27 (5.2 × 0.62 cm, Bio-Rad) stationary phase eluted with 25 mM sodium citrate, 1 mM potassium perchlorate, 0.3 mM sodium azide, pH 7.5 at a flow rate of 0.7 ml min^{-1} (22); reverse-phase data obtained with μBondapak C_{18} (30 × 0.4 cm, Waters) eluted with a gradient of 0.2 M KH_2PO_4 pH 4 at initial eluent, and 0.2 M KH_2PO_4 pH 4.0 containing 20% methanol as final gradient at a flow rate of 2 ml min^{-1} (23); paired-ion data obtained with μBondapak C_{18} (30 × 0.4 cm, Waters) eluted in a presence of a tetrabutylammonium counterion, with a linear gradient of 0.1 M KH_2PO_4 pH 6.4 containing 5 mM PIC reagent A (Waters) initial eluent and 0.1 M KH_2PO_4 containing 5 mM PIC reagent A and 30% methanol as final eluent, at a flow rate of 2 ml min^{-1} (23).

Table 4. Ultraviolet absorbance spectra characteristics of cyclic nucleotides

	pH 1		pH 7		pH 11	
	λ_{max}	$\varepsilon \times 10^{-3}$	λ_{max}	$\varepsilon \times 10^{-3}$	λ_{max}	$\varepsilon \times 10^{-3}$
cAMP	257	14.5	261	15.4	261	15.4
cGMP	256	12.3	252	13.5	258	11.4
cIMP	247	11.8	—	—	252	12.8
cCMP	279	12.4	272	9.3	271	9.2
cUMP	262	9.9	262	9.9	261	7.3
cdTMP	—	—	261	9.9	260	7.7

Data compiled from refs 3, 24 and 25, and from unpublished data from the author's laboratory.

teristic physical properties. Ultraviolet absorbance spectra readily confirm the identity of the base and, for a homogeneous cyclic nucleotide preparation, provide a method of estimation from the molar absorbance coefficients; u.v. absorbance data will not, however, differentiate cyclic and non-cyclic nucleotides nor cyclic nucleotide isomers, and possess the additional complication of being pH-dependent. Thus u.v. absorbance spectra must be obtained under precisely predetermined acid, neutral or alkaline conditions (*Table 4*), taking into account that unbuffered cyclic nucleotide solutions are themselves acidic; such spectra can only be used for identification in combination with an analytical chromatographic system.

Mass spectrometric analysis does provide a means of unequivocal cyclic nucleotide identification, readily distinguishing 3′,5′- and 2′,3′-isomers. While chemical ionization and electron impact mass spectra can only be obtained from derivatized samples and contain only weak molecular ions, fast atom bombardment mass spectra can be obtained from the underivatized cyclic nucleotides in a glycerol–water matrix, and contain strong peaks corresponding to the protonated molecular ion, its sodium, disodium, glycerol and sodium plus glycerol adducts (*Table 5*). The presence of these major peaks confirms the identity of a cyclic nucleotide;

Table 5. Fast atom bombardment mass spectra of cyclic nucleotides

Cyclic nucleotides	M_r	MH^+	$[M+Na]^+$	$[MNa+Na]^+$	$[M+Gro+H]^+$	$[M+Gro+Na]^+$
cAMP	329	330[a]	352	—	422	—
cGMP	345	346	368	390[a]	438	460
cIMP	330	331	353[a]	375	423	445
cCMP	305	306[a]	328	—	398	—
cUMP	306	307	329[a]	351	399	421
cdTMP	304	305[a]	327	349	397	—

m/z values of intense peaks arising from sample-derived ions.
[a] Represents most intense peaks obtained in spectra of analytes in tissue extracts.
Data obtained from refs 26, 27, 28 and unpublished data from the author's laboratory.

Table 6. Characteristic fragments obtained from protonated molecular ions of cyclic nucleotides

Cyclic nucleotide	$[BH_2]^+$	$[BH+28]^+$	$[BH+42]^+$
cAMP	136	164	178
cGMP	152	180	194
cIMP	137	165	179
cCMP	112	140	154
cUMP	113	141	155
cdTMP	111	139	—

m/z values of diagnostic peaks obtained from fragmentation of protonated molecular ions of 3′,5′-cyclic nucleotides; fragmentation of the sodium adduct of protonated molecular ion will yield these peaks together with more intense peaks at m/z values increased by 22 mass units. cdTMP does not yield the $[BH+42]^+$ fragment due to absence of 2′-O in its structure (26, 27, 28).

application of tandem mass spectrometry or mass-analyzed ion kinetic energy spectrum scanning to observe the fragmentation of the protonated molecular ion enables the differentiation of 2′,3′- and 3′,5′-cyclic nucleotide isomers (26, 27, 28). The 3′,5′-cyclic nucleotide spectra contain three strong peaks corresponding to protonated base, protonated base plus 28 mass units and protonated base plus 42 mass units, while the latter is virtually absent from spectra of 2′,3′-isomers (*Table 6*).

Proton-NMR yields useful spectra of cAMP, but for other 3′,5′- cyclic nucleotides tight spins often make spectra unreliable and irreproducible (29). Carbon-13 NMR, however, provides spectra with characteristic shifts that can be used diagnostically, since those obtained from the 3′,5′-cyclic nucleotides (30, *Table 7*) differ significantly from those of the 2′,3′-cyclic nucleotides (31).

Table 7. Carbon 13-NMR data for 3,5′-cyclic nucleotides

	δ p.p.m. downfield from TMS signal				
	cAMP	**cGMP**	**cCMP**	**cUMP**	**cdTMP**
1′	92.69	93.20	96.18	95.67	86.53
2′	73.47	73.27	73.37	73.03	35.71
3′	72.91	73.03	72.91	72.90	76.01
4′	78.48	78.30	78.28	78.16	77.48
5′	68.54	68.60	68.50	68.32	68.40
2	153.82	154.89	158.18	152.22	152.51
4	149.05	152.23	167.53	167.21	167.33
5	119.52	117.65	97.38	103.42	112.96
6	156.50	159.86	143.08	143.36	138.91
8	140.78	139.19	—	—	—

Data reproduced from Lapper, R.D., Mantsch, H.H. and Smith, I.C.P. (1973) *J. Am. Chem. Soc.*, **95**, 2878–2880, with permission.

5. ESTIMATION OF CYCLIC NUCLEOTIDES

While cyclic nucleotide concentrations in chromatographically purified extracts can be estimated by spectrophotometry (3, 24), luciferin–luciferase activation (32) or enzymatic cycling (33), less highly purified cyclic–nucleotide-containing extracts are more usually analyzed by one of the three following methods.

5.1. Radioimmunoassay

Radioimmunoassay (RIA) can be carried out using commercially available kits (Amersham International, New England Nuclear or Biogenesis) or from components produced in-house. For the latter, antisera are raised against the cyclic nucleotide coupled to a carrier protein, usually human or bovine serum albumin, bovine thyroglobulin, keyhole limpet hemocyanin or poly-L-lysine, using a carbodiimide as coupling agent (34, 35); a high-specific-activity radiolabeled antigen is produced by synthesizing the 2′-*O*-succinyl tyrosinyl methyl ester derivative of the cyclic nucleotide and radioiodinating it with 40–70 MBq (1–2 mCi) sodium [^{125}I]iodide (34, 35). Conventionally, assays are carried out in a total buffered volume of 100–700 μl containing standard or unknown cyclic nucleotide, $1-2 \times 10^4$ d.p.m. (0.2–0.3 kBq) of the radiolabeled cyclic nucleotide derivative, sufficient antibody to bind 33–50% of the latter, and a γ-globulin protein as carrier (34, 35, 36). After incubation at 4°C to equilibrium, usually 12–20 h, separation of bound and free antigen can be carried out by fractional precipitation (34), filtration (34, 36) or addition of a second antibody (35), and the resultant solution counted in a γ-counter. If required, the sensitivity of the assay can be increased by up to two orders of magnitude by acetylating the cyclic nucleotide standards and unknowns (35): it is then possible to determine cyclic nucleotides in the 10^{-15}–10^{-16} mole range. Extraction blanks, in which buffer instead of tissue is put through the purification procedure, and extraction controls, in which the purified extract is treated with cyclic nucleotide phosphodiesterase, are required to detect any cross-reactivity due to non-specificity of the antisera, in addition to the requisite controls to determine background counts, background binding and total binding.

5.2. Binding protein saturation analysis

Binding protein saturation analysis is based upon the same principle as radioimmunoassay, that is competition between labeled and unlabeled cyclic nucleotides for a constant, limited number of available binding sites on a macromolecule that selectively binds them (37, 38). In this case, rather than the antisera employed in RIA, naturally occurring proteins are used, cAMP-binding protein being extracted from bovine skeletal muscle (39) or adrenal glands (37) and cGMP-binding protein from lobster tail muscle (40). No analogous assays are currently available for the other cyclic nucleotides.

Assays are carried out in a total buffered volume of 50–300 μl containing $15-30 \times 10^3$ d.p.m. (0.3–0.5 kBq) ^{3}H-labeled cyclic nucleotide, 0.2–40 pmol of standard or unknown unlabeled cyclic nucleotide, and sufficient binding protein to bind 20–30% of the [^{3}H]cAMP. Separation of bound and free cyclic nucleotide is carried out by membrane filtration (38) or charcoal adsorption (37) prior to counting in a scintillation counter. Analogous controls to those described for radioimmunoassay are required. The disadvantage of a lower sensitivity for the binding saturation method compared to the radioimmunoassay may be offset by the advantages of a much longer half-life of assay components and the avoidance of handling a γ-emitting radioisotope.

5.3. Protein kinase assay

The protein kinase assay is based upon the reaction:

$$[\gamma\text{-}^{32}\text{P}]\text{ATP} + \text{protein} \xrightarrow[\text{protein kinase}]{\text{cNMP}} [^{32}\text{P}]\text{protein} + \text{ADP}.$$

The protein kinase employed is cAMP- or cGMP-dependent, and at saturating levels of [γ-^{32}P]ATP and protein substrate the rate of phosphorylation of the protein is directly proportional to the cyclic nucleotide concentration. cAMP-dependent protein kinase can be prepared from bovine heart or skeletal muscle by well-established procedures (41, 42), and cGMP-dependent protein kinase from lobster tail muscle (40); alternatively kinases can be purchased commercially (Sigma Chemical Co.). Incubation is carried out in a total volume of 100–200 μl, containing $1-2 \times 10^6$ d.p.m. (30–60 kBq) [γ-^{32}P]ATP and 30–50 μg of histone or casein as protein substrate, with a stopping solution of a protein precipitant used to terminate the reaction. The separation step can be either centrifugation and aspiration, or protein adsorption on to filter-paper disks, prior to counting of radiolabeled protein (40, 41, 42). A plot of radioactivity incorporated against cyclic nucleotide added should yield a linear relationship after subtraction of a control containing all components except active protein kinase, and permit cyclic nucleotide assay in the 10^{-13}–10^{-14} mol range. The method is more sensitive to high ionic content than the above two alternatives, and thus should not be used with KOH-neutralized perchloric acid extracts.

6. ENDOGENOUS CONCENTRATIONS OF CYCLIC NUCLEOTIDES

Cyclic nucleotide concentrations vary naturally, dependent upon the metabolic state of the tissue under analysis. A representative selection of resting-state tissue-level ranges is shown in *Table 8*, to provide some guidance for sampling size for tissue extraction prior to estimation by one of the methods outlined in Section 5. As a generalization, cGMP levels in mammalian tissues are tenfold to 100-fold lower than the levels of cAMP, with cCMP, cUMP, cIMP and cdTMP an order of magnitude or more lower than those of cGMP. Some tissues, for example brain, testis and lung, are enriched in cGMP, while cGMP levels in plasma and urine are approximately half those of cAMP. In plants the levels of cyclic nucleotides are lower than in mammals, but the ratio of cyclic nucleotide to nucleotide triphosphate concentration is similar (43).

Table 8. Endogenous concentrations of cyclic nucleotides

Tissue	Cyclic nucleotide concentration (pmol/g tissue)		
	cAMP	cGMP	cCMP
Liver	650–1200	11–19	1–8
Kidney	720–1100	29–47	0.5–9
Pituitary	620–1400	7–8	—
Heart	590–990	12–24	1–4.6
Fat	30–82	2–6	—
Lung	940–1380	52–158	1–5

Data taken from refs 33, 34, 37, 38, 40, 41 and from data from author's own laboratory.

7. PHYSIOLOGICAL EFFECTS OF CYCLIC AMP

At least in mammalian cells, most metabolic pathways contain one or more enzymes that act as a substrate for cAMP-dependent protein kinase, resulting in a change in the enzyme's activity and, consequently, in the flux through the pathway. *Table 9* lists a number of such

Table 9. Pathways sensitive to cAMP-responsive phosphorylation

Pathway/process	Enzyme phosphorylated	Pathway/process	Enzyme phosphorylated
Carbohydrate metabolism	Phosphorylase kinase Phosphofructokinase Glycogen synthetase Fructose 1,6-bisphosphatase Pyruvate kinase	Lipid metabolism	Acetyl CoA carboxylase ATP-citrate lyase Cholesterol esterase Glycerol 3-phosphate acyl transferase Phospholipid methyl-transferase Lipase
Protein metabolism	Actin C protein Factor-3 Filamin Myelin Phenylalanine monoxygenase	Muscle contraction	Troponin 1 Myosin light chain Myosin light chain kinase
		Na^+-K^+ ATPase	

enzymes established in carbohydrate, lipid and protein metabolism and in the regulation of muscle contraction and of Na^+-K^+ ATPase activity; there are more comprehensive reviews of mammalian cAMP-dependent protein kinase properties and substrates (44–46) and of plant cAMP-sensitive proteins (43).

8. DERIVATIVES OF CYCLIC NUCLEOTIDES

Cyclic nucleotides do not readily permeate cell membranes; thus in order to observe the effects of a cyclic nucleotide upon an organism, intact cell or tissue, a more lipophilic derivative is employed. Compounds such as N^6,2′-*O*-dibutyryl-cAMP and N^8-bromo-cGMP (*Figure 4*) are commercially available and used routinely. Reviews of the range of such derivatives and methods of synthesis should be consulted (25, 47–50). Whatever derivative is selected for testing, it is essential that adequate controls are utilized; for example, in the case of dibutyryl cAMP, butyric acid controls must be used to ensure that any observed effects are not the result of butyryl group release.

9. ESTIMATION OF NUCLEOTIDE CYCLASE ACTIVITY

Assays of nucleotide activity may be carried out with radiolabeled or unlabeled substrate. With unlabeled substrate, the cyclic nucleotide formed is assayed by one of the methods detailed in Section 5, but for assays of large numbers of samples, use of radiolabeled substrate is logistically preferable. With adenylyl cyclase, most assay protocols involve radiolabeled ATP as the substrate, a separation of the cAMP product from other labeled assay reactants and side-products, then determination of radiolabel in this product. [^{3}H]- or [^{14}C]-ATP, despite the advantage of longer half-lives than [^{32}P]-ATP, have lower specific activities and label the potential side-products adenosine and adenine; thus, where practicable, [α-^{32}P]-ATP is the preferred choice for substrate. In addition to the inclusion of essential cyclase cofactors (usually at least one inorganic salt), the composition of the assay incubation medium is designed to reduce consumption of the nucleotide triphosphate by non-cyclase

Figure 4. Cell-permeating cyclic nucleotide derivatives.

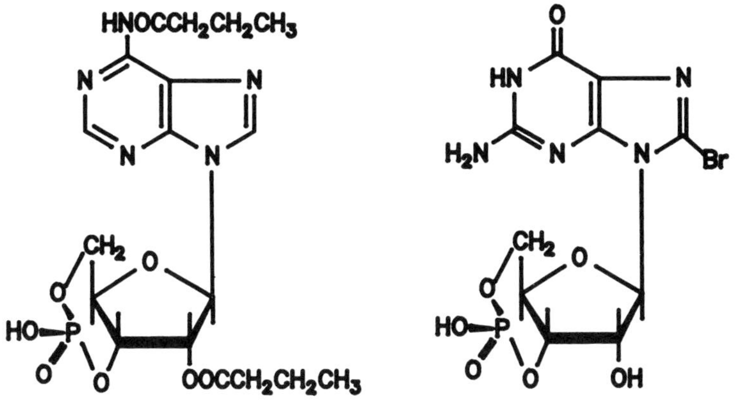

N6,2'-O-dibutyryl cyclic AMP

8-Bromo-cyclic GMP

reactions, and to protect the nascent cyclic nucleotide product from degradation. For the latter, cyclic nucleotide phosphodiesterase activity can be minimized by the inclusion of an inhibitor such as an alkylxanthine or papaverine, and/or by inclusion of unlabeled cyclic nucleotide 'carrier'. For the former, either the use of a radiolabeled nucleoside triphosphate (NTP) imidoderivative (*Figure 5*), a good cyclase but poor NTPase substrate, or the inclusion of an NTP-regenerating system, either creatine kinase plus creatine phosphate or pyruvate kinase plus phosphoenolpyruvate, is routinely used (51, 52).

Incubation volumes normally vary from 50 to 500 μl, containing $2-4 \times 10^5$ d.p.m. (3–6 kBq) of labeled substrate; termination of the reaction can be carried out by boiling, acid precipitation or use of a stopping solution, such as zinc acetate/Na_2CO_3, which selectively precipitates non-cyclic nucleotides (52). Separation of the product is carried out by sequential chromatography on short columns of Dowex 50, H^+ form, and neutral alumina (51, 52) PEI cellulose (51, 52) or QAE Sephadex and alumina (53), prior to determination of radiolabel

Figure 5. Adenylyl-imidodiphosphate (AMP–PNP).

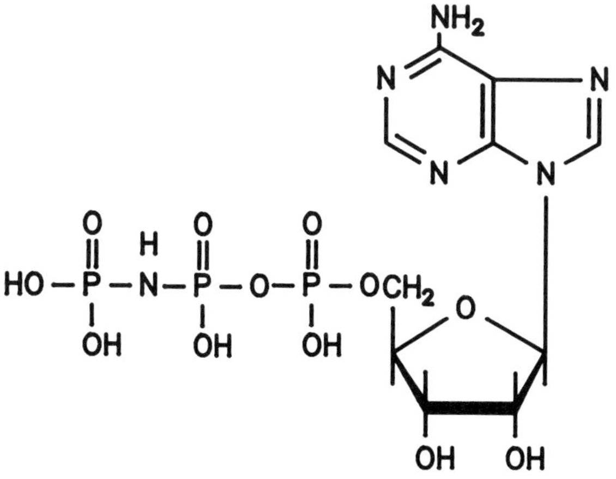

present. Particularly in the case of adenylyl cyclase, treatment of membranous cyclase preparations by gentle detergents or solubilizers may increase the apparent activity, but care must be taken to ensure that this does not destroy sensitivity to agonists; the variation of cyclase preparatory procedures depends upon the tissue under consideration (54) and cyclase sensitivity to agonists is also tissue dependent (8).

10. ESTIMATION OF CYCLIC NUCLEOTIDE PHOSPHODIESTERASE ACTIVITY

Cyclic nucleotide phosphodiesterase activity exists as multiple phosphodiesterase isoenzymes, many of which have been purified to homogeneity, and which can be grouped into at least five different isoenzyme families: calcium-calmodulin dependent, cGMP-stimulated, cGMP-inhibited, cAMP-specific and cGMP-specific phosphodiesterases (55). Purification of an individual phosphodiesterase is dependent upon the particular phosphodiesterase selected and tissue source, but usually employs conventional protein purification chromatography methods, including use of affinity chromatography with a cyclic nucleotide or calmodulin ligand (56).

The most sensitive and reproducible assays of phosphodiesterase activity employ a radiolabeled substrate, although alternative assays with fluorescent derivatives (57) or colorimetric phosphate determination (58) are also used. In the radiolabeled assays, $5-8\times10^4$ d.p.m. (0.8–1.2 kBq) of ^{3}H-labeled cyclic nucleotide are incubated with the enzyme preparation in a buffered total volume of 100–500 μl containing the requisite cofactors. Termination of the reaction is accomplished by boiling or by acid precipitation. The mononucleotide product is not readily separated from the cyclic nucleotide substrate, therefore in most assays an auxiliary enzyme, capable of hydrolyzing mononucleotide to nucleoside, is incubated in excess with the supernatant from the primary incubation; this secondary enzyme being either partially purified snake venom (*Crotalus adamanteus*, *Ophiophagus hannah* or *Crotalus durissus terrificus*) or purified nucleotidase. At the end of this second stage the reaction is again terminated by heating or acid precipitation, prior to chromatographic separation. The most practicable separation is accomplished using short columns of anion-exchange Sephadex, DEAE 25 or QAE 25, eluted by H_2O or 30 mM ammonium formate, or single ion-exchanger, AG1 or Dowex 1, eluted by methanol. The nucleoside end-product is eluted readily while the phosphorylated compounds are retained. After separation the radiolabeled end-product is determined by scintillation counting (59).

11. ESTIMATION OF CYCLIC NUCLEOTIDE-DEPENDENT PROTEIN KINASE ACTIVITY

Cyclic nucleotide-sensitive and -dependent protein kinase can be soluble or particulate (60). Assay of cyclic nucleotide-dependent protein kinase in tissue extracts is based upon the principle elaborated in Section 5, with the incorporation of radiolabel from [γ-^{32}P]ATP into a kinase substrate protein being determined (40–42). Endogenous proteins may be utilized as the kinase substrate; credible demonstration of cyclic nucleotide-dependent protein kinase activity must include a dose-dependency upon the cyclic nucleotide concentration.

12. CYCLIC NUCLEOTIDE IMMUNOCYTOCHEMISTRY

Sections 2–11 consider the measurement of cyclic nucleotides and related enzymes in tissue extracts, broken cell preparations and purified analytes. The methodology also exists for immunocytological examination of intact cells and organelles, with effective analysis of cyclic nucleotides and cyclic nucleotide-dependent protein kinase established (61–65) using antibodies generated against cyclic nucleotide conjugates (as described for RIA in Section 5) and against the protein kinase holoenzyme.

13. REFERENCES

1. Sundralingham, M. (1975) *Ann. NY Acad. Sci.*, **255**, 3.

2. Rall, T.W. and Sutherland, E.W. (1958) *J. Biol. Chem.*, **232**, 1065.

3. Ashman, D.F., Lipton, R., Melicow, M.M. and Price, T.D. (1963) *Biochem. Biophys. Res. Commun.*, **11**, 330.

4. Newton, R.P., Salih, S.G., Salvage, B.J. and Kingston, E.E. (1984) *Biochem. J.*, **221**, 665.

5. Newton, R.P., Kingston, E.E., Hakeem, N.A., Salih, S.G., Beynon, J.H. and Moyse, C.D. (1986) *Biochem. J.*, **236**, 431.

6. Newton, R.P., Chiatante, D., Ghosh, D., Brenton, A.G., Walton, T.J., Harris, F.M. and Brown, E.G. (1989) *Phytochem.*, **28**, 2243.

7. Ishiyama, J. (1975) *Biochem. Biophys. Res. Commun.*, **65**, 286.

8. Herman, R.H. and Taunton, O.D. (1980) in *Principles of Metabolic Control in Mammalian Systems* (R.H. Herman, R.M. Cohn and P.D. McNamara eds). Plenum Press, New York, p. 424.

9. Birnbaumer, L. and Iyengar, R. (1982) in *Handbook of Experimental Pharmacology* (J.A. Nathanson and J. W. Kebabian eds). Springer Verlag, Berlin, Vol. 58, Part I, p. 153.

10. Bitenski, M.W., Wheeler, M.A., Rasenick, M.M., Yamazaki, A., Stein, P.J., Halliday, K.R. and Wheeler, J.L. (1984) *Proc. Natl. Acad. Sci. USA*, **79**, 3408.

11. Tremblay, J., Gerzer, R.M., Vinay, P., Pang, S.C., Beliveau, R. and Hamet, P. (1985) *FEBS Lett.*, **181**, 17.

12. Anderson, T.R. (1982) *Mol. Cell Endocr.*, **28**, 373.

13. Wollenberger, A., Ristau, O. and Schoffa, G. (1960) *Pfleugers Arch. Gesamte Physiol. Meschen Tiere*, **270**, 399.

14. Mayer, S.E., Stull, J.T. and Wastila, W.B. (1974) *Methods Enzymol.*, **38**, 3.

15. Brown, E.G. (1991) *Methods in Plant Biochemistry*, **5**, 53.

16. Schultz, G., Böhme, E. and Hardman, J.G. (1974) *Methods in Enzymol.*, **38**, 9.

17. Murad, F., Manganiello, V. and Vaughan, M. (1971) *Proc. Natl. Acad. Sci. USA*, **68**, 736.

18. Scavennec, J., Carcassonne, Y., Gastaut, J., Blanc, A. and Cailla, H.L. (1981) *Cancer Res.*, **41**, 3222.

19. White, A.A. and Zenser, T.V. (1971) *Anal. Biochem.*, **41**, 372.

20. Davis, C.W. and Daly, J.W. (1979) *J. Cyclic Nucleotide Res.*, **5**, 65.

21. Böhme, E. and Schultz, G. (1974) *Methods Enzymol.*, **38**, 27.

22. Khym, J.X. (1978) *J. Chromatog.*, **151**, 421.

23. Brown, E.G., Newton, R.P. and Shaw, N.M. (1982) *Anal. Biochem.*, **123**, 378.

24. Smith, M., Drummond, G.I. and Khorana, H.G. (1961) *J. Am. Chem. Soc.*, **83**, 698.

25. Revankar, G.R. and Robins, R.K. (1982) in *Handbook of Experimental Pharmacology*, (J.A. Nathanson and J.W. Kebabian, eds). Springer Verlag, Berlin, Vol. 58, Part I, p. 16.

26. Kingston, E.E., Beynon, J.H. and Newton, R.P. (1984) *Biomed. Mass Spectrom.*, **11**, 367.

27. Kingston, E.E., Beynon, J.H., Newton, R.P. and Liehr, J.G. (1985) *Biomed. Mass Spectrom.*, **12**, 525.

28. Newton, R.P., Brenton, A.G., Ghosh, D., Walton, T.J., Langridge, J.I., Harris, F.M. and Evans, A.E. (1991) *Anal. Chim. Acta.*, **247**, 161.

29. Blackburn, B.J., Lapper, R.D. and Smith, I.C.P. (1973) *J. Am. Chem. Soc.*, **95**, 2873.

30. Lapper, R.D., Mantsch, H.H. and Smith, I.C.P. (1973) *J. Am. Chem. Soc.*, **95**, 2878.

31. Lapper, R.D. and Smith, I.C.P. (1973) *J. Am. Chem. Soc.*, **95**, 2880.

32. Sutherland, C.A. and Johnson, R.A. (1974) *Methods Enzymol.*, **38**, 62.

33. Goldberg, N.D., O'Toole, A.G.O. and Haddox, M.K. (1972) *Adv. Cyclic Nucleotide Res.*, **2**, 81.

34. Steiner, A.L. (1974) *Methods Enzymol.*, **38**, 96.

35. Delaage, M.A., Roux, D. and Cailla, H.L. (1978) in *Molecular Biology and Pharmacology of Cyclic Nucleotides* (G. Folco and R. Paoletti eds). Elsevier/North Holland, Amsterdam, p. 151.

36. Brooker, G., Harper, J.F., Terasaki, W.L. and Maylan, A. (1979) *Adv. Cyclic Nucleotide Res.*, **10**, 1.

37. Brown, B.L., Ekins, R.P. and Albano, R.D.M. (1972) *Adv. Cyclic Nucleotide Res.*, **2**, 25.

38. Gilman, A.G. and Murad, F. (1974) *Methods Enzymol.*, **38**, 49.

39. Miyamoto, E., Kuo, J.F. and Greengard, P. (1969) *J. Biol. Chem.*, **244**, 6395.

40. Kuo, J.F. and Greengard, P. (1970) *J. Biol. Chem.*, **245**, 2493.

41. Kuo, J.F. and Greengard, P. (1972) *Adv. Cyclic Nucleotide Res.*, **2**, 41.

42. Rannels, S.R., Beasley, A. and Corbin, J.D. (1982) *Methods Enzymol.*, **99**, 55.

43. Newton, R.P. and Brown, E.G. (1987) in *Hormones, Receptors and Cellular Interactions in Plants* (C.M. Chadwick and D.R. Garrod eds). Cambridge University Press, Cambridge, p. 115.

44. Beavo, J.A. and Mumby, M.C. (1982) in *Handbook of Experimental Pharmacology* (J.A. Nathanson and J.W. Kebabian eds). Springer-Verlag, Berlin, Vol. 58, Part I, p. 363.

45. Kuo, J.F. and Shoji, M. (1982) in *Handbook of Experimental Pharmacology* (J.A. Nathanson and J.W. Kebabian eds). Springer-Verlag, Berlin, Vol. 58, Part I, p. 393.

46. Shackter, E., Stadtman, R., Jurgensen, S.R. and Chock P. Boon (1988) *Methods Enzymol.*, **159**, 3.

47. Pasternak, T.H. and Weiman, G. (1974) *Methods Enzymol.*, **38**, 399.

48. Miller, J.P. (1977) in *Cyclic 3′,5′-Nucleotides: Mechanisms of Action* (H. Cramer and J. Schultz eds). J. Wiley and Sons, London, p. 77.

49. Pasternak, T.H. (1979) in *Cyclic Nucleotides and Therapeutic Perspectives* (G. Cehovic and G.A. Robison eds). Pergamon Press, Oxford, p. 4.

50. Botelho, L.H.P., Rothermel, J.D., Coombs, R.V. and Jastorff, B. (1988) *Methods Enzymol.*, **159**, 159.

51. Schultz, G. (1974) *Methods Enzymol.*, **38**, 115.

52. Johnson, R.A. and Salomon, Y. (1991) *Methods Enzymol.*, **195**, 3.

53. Newton, R.P., Salvage, B.J. and Hakeem, N.A. (1990) *Biochem. J.*, **265**, 581.

54. Johnson, R.A. and Corbin, J.D. (1991) *Methods Enzymol.*, **195** (Section IC), p. 65; and **195** (Section IIIB), p. 355.

55. Beavo, J. (1990) in *Cyclic Nucleotide Phosphodiesterases: Structure, Regulation and Drug Action* (J. Beavo and M.D. Houslay eds). J. Wiley and Sons, Chichester, p .3.

56. Corbin, J.D. and Johnson, R.A. (1988) *Methods Enzymol.*, **159** (Section VI), p. 543.

57. Secrist, J.A. (1974) *Methods Enzymol.*, **38**, 428.

58. Baykov, A.A., Evtushenko, O.A. and Avaeva, S.M. (1988) *Anal. Biochem.*, **171**, 266.

59. Kincaid, R.L. and Manganiello, V.C. (1991) *Methods Enzymol.*, **195**, 457.

60. Rudolf, S.A. and Keueger, B.K. (1979) *Adv. Cyclic Nucleotide Res.*, **10**, 107.

61. Spruill, W.A. and Steiner, A.L. (1979) *Adv. Cyclic Nucleotide Res.*, **10**, 169.

62. Kapoor, C.L. and Steiner, A.L. (1982) in *Handbook of Experimental Pharmacology* (J.A. Nathanson and J.W. Kebabian eds). Springer-Verlag, Berlin, Vol. 58, Part I, p. 333.

63. Jungmann, R.A., Kuettel, M.R. Squinto, S.P. and Kwast-Welfeld, J. (1988) *Methods Enzymol.*, **159**, 225.

64. Byus, C.V. and Fletcher, W.H. (1988) *Methods Enzymol.*, **159**, 236.

65. Fletcher, W.H., Ishida, T.A., Van Patten, S.M. and Walsh, D.A. (1988) *Methods Enzymol.*, **159**, 255.

CHAPTER 11
STEROID HORMONES

R. E. Leake

1. INTRODUCTION

Hormones are blood-borne messengers that have specific target cells. A target cell for a hormone is defined by the possession of receptors (usually several thousand per cell) which show both specificity and high affinity for the particular hormone. The steroid hormones, in common with thyroid hormone, have their receptors inside the target cell, whereas polypeptide hormones have receptors on the cell surface. The chemical basis for this is simple in that all steroid hormones are derived from the parent molecule cholesterol (a molecule containing 27 carbon atoms and so called a C27 molecule). *Figure 1* outlines the synthetic pathways for steroid hormones. Steroid hormones are membrane soluble, so uptake into the cell occurs down a concentration gradient. Thus, an adequate concentration of high-affinity receptors within a cell should be able to pull steroid into the cell from the carrier proteins in the bloodstream. One of the early experiments (1) showed that target cells (receptor-containing cells) can concentrate steroid relative to plasma levels. The steroid hormones act by binding and activating the intracellular receptors such that the hormone–receptor complex can interact with specific sites on the genome, either up- or down-regulating the transcription of appropriate genes. Because steroid hormones act (to a large extent) by regulating gene expression, the responses that they induce in their target cells tend to be long-term responses. These responses include induction of specific enzymes, increased DNA synthesis and cell growth and regulation of sexual activity.

Steroid hormones can be subdivided on the basis of their chemistry and their biological function. However, some steroids, particularly synthetic molecules, have relatively similar binding affinities for two different classes of steroid receptors and so can have two different types of effect, for example they can act as both progestins and glucocorticoids. Similarly, molecules which are chemically not steroid can, by the nature of their high affinity for steroid receptors, demonstrate either steroid or anti-steroid activities. This property has been greatly exploited in the treatment of several steroid-based diseases and so these molecules will be included in this review.

Steroid hormones are divided into five general biological subgroups. These are progesterone (and progestins, molecules with similar biological properties), mineralocorticoids, glucocorticoids, androgens and estrogens. The first three subgroups are all C21 molecules, while the androgens are C19 and the estrogens C18.

The synthesis of steroid hormones has been reviewed elsewhere (2, 3).

2. PLASMA BINDING PROTEINS FOR STEROIDS

Steroid hormones are inactivated rapidly by the liver, by convertion to glucuronides, and are then excreted via the kidney. In order to protect them prior to their exerting physiological effects in the target tissue(s), they are transported in the blood bound to various transport

Figure 1. Principal pathways of steroid synthesis in the adrenal cortex. The enzymes involved are: 1, cholesterol side-chain cleavage complex; 2, 17-hydroxylase; 3,3β-hydroxysteroid dehydrogenase; 4, C17,20-desmolase; 5, 21-hydroxylase; 6,11β-hydroxylase; 7, 18-hydroxylase; 8, 18-hydroxysteroid dehydrogenase; 9, sulfotransferase; 10, sulfatase.

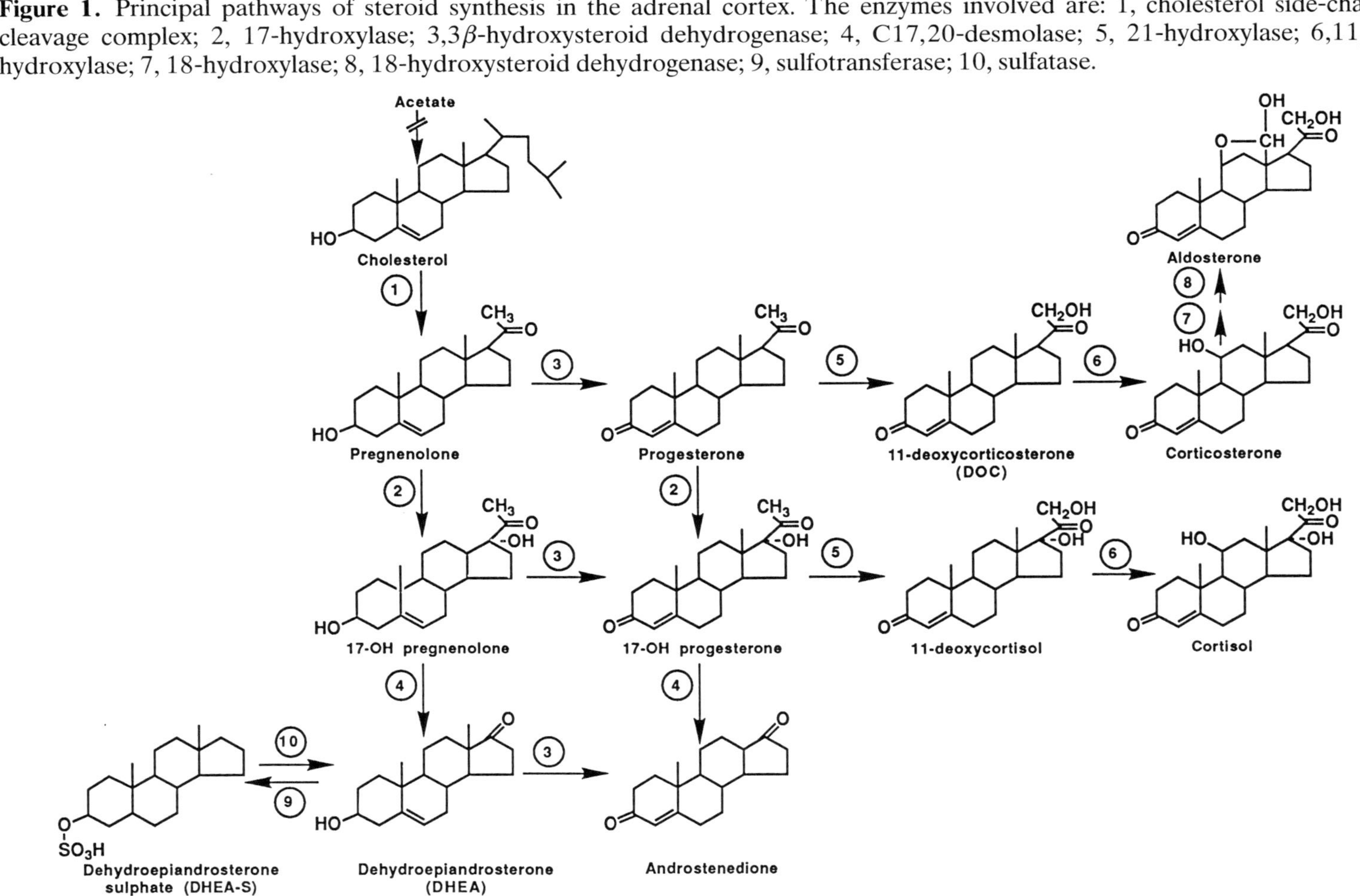

proteins. The transport proteins may also hold a reservoir of steroid in times of shortage. The plasma transport proteins are subdivided into two groups, those with specific, high affinity for particular steroids (cortisol-binding globulin, or CBG; and sex-hormone-binding globulin, or SHBG) and those with non-specific, low-affinity but high-capacity binding, such as albumin.

Cortisol-binding globulin (CBG) is an α-glycoprotein with molecular weight of 500 kDa. It binds the following steroids with high affinity: cortisol, corticosterone, 11-deoxycortisol, progesterone and 17-OH progesterone. The plasma concentration of CBG is about 30 mg l^{-1}. This is about three times the plasma concentration of cortisol, and so free cortisol levels are low, except in clinical cases of excessive cortisol secretion.

Sex-hormone-binding globulin (SHBG) is a β-glycoprotein and a dimer. It has a molecular weight of about 42 kDa. It binds dihydrotestosterone most strongly, followed by testosterone and estradiol. Levels of SHBG are reduced (normal plasma level about 3 mg l^{-1}) in clinical cases of acne in men and acute masculinization in postmenopausal women.

There are several species differences in plasma binding proteins for steroids. Perhaps the most important example is that estrogens are bound by α-fetoprotein in rats but not in humans. Conversely, rats do not have SHBG.

At any one time, the majority of steroid is bound to carrier protein(s). However, the rate of uptake into tissues is controlled by the concentration of free steroid. Hence the concentration of plasma binding protein can regulate the activity of the steroid within its target cell. However, this is not a direct relationship and questions remain about the precise role of the plasma binding proteins. For example, it has been suggested that SHBG might increase, rather than decrease, the ability of estrogen to enter breast cells. A classic paper summarizes steroid–plasma protein interactions (4).

Cortisol is bound principally to CBG and there is virtually no binding to albumin at normal plasma levels of cortisol. Testosterone is bound principally to SHBG, but also to CBG and albumin. Estradiol is bound to SHBG and albumin. Progesterone is bound principally to CGG but also to albumin. Aldosterone binds with low affinity to both CBG and albumin but largely circulates in the unbound form.

3. BIOLOGICAL PROPERTIES OF STEROIDS

The different classes of steroid have many biological effects, ranging from metabolic to behavioral. Of necessity, the following review is selective.

3.1. Glucocorticoids

One of the major roles of glucocorticoids is to stimulate glucose supply. The glucose-exporting tissues of the body are the liver and the kidney cortex. Glucocorticoids promote hepatic glucose output by inducing the synthesis (at gene level) of the rate-limiting enzymes of gluconeogenesis, increasing the supply from peripheral tissues of appropriate carbon skeletons (amino acids) and combining with other glucose-sparing hormones (e.g. inducing synthesis of growth hormone receptor so that growth hormone will increase lipolysis to generate energy). One of the best-studied enzymes under glucocorticoid control is hepatic tyrosine aminotransferase (TAT), an enzyme required for the mobilization of the carbon skeleton of tyrosine.

Glucocorticoid action regulates blood pressure, with excess glucocorticoid resulting in hypertension. Glucocorticoids also have much less clearly defined activities in relation to a whole variety of tissues (almost all tissues seem to contain glucocorticoid receptors).

Adequate supplies of active glucocorticoid are essential to combat stress. They are also known to have anti-inflammatory effects and anti-allergic effects. For example, changes have been noted in glucocorticoid–receptor interaction in asthma patients resistant to glucocorticoid therapy, compared to responders (5). A similar role for glucocorticoid receptors has been identified in the fibroblasts taken from synovial tissue of patients suffering from rheumatoid arthritis in the knee joint (6).

3.2. Mineralocorticoids

The mineralocorticoids act to conserve sodium and excrete potassium and, in so doing, regulate blood pressure — excess mineralocorticoid leading to hypertension. Under normal circumstances, aldosterone promotes the reabsorption of sodium ions and the corresponding loss of potassium or hydrogen ions. This occurs in the distal tubule cells of the kidney. If the body becomes sodium depleted for any reason; renin is released by the kidney. The renin then activates angiotensinogen to give, first, angiotensin-I, then angiotensin-II, which acts directly on the zona glomerulosa of the adrenal to promote synthesis and release of aldosterone.

3.3. Estrogens

Estrogens regulate the growth of the female sexual tissues and control the secondary sex characteristics. In humans, the prime role of the estrogens is, perhaps, to induce synthesis of the cells required to create the new lining of uterine lumen after the end of menses. Estrogens have less easily observed effects on other target tissues, such as the cervix, vagina and breast. However, a role for estrogens in the etiology of breast cancer has long been suspected. More recent evidence (7) would suggest that this role may be played by locally synthesized estrogen, rather than by plasma steroid. In most tissues, though not the ovaries, one of the main effects of estrogen is to induce the synthesis of the progesterone receptor, in order to prepare the tissue to repond to progesterone in the second half of the menstrual cycle. Comparison of the response of breast and endometrial epithelium illustrates the contrasting effects of estrogen and progesterone. In the endometrium, all the DNA syntheis takes place in the proliferative phase of the cycle and progesterone rapidly inhibits the estrogen-induced DNA synthesis. In the breast, the epithelium undergoes maximum DNA synthesis in the secretory phase of the cycle, under the stimulation of both estrogen and progesterone (8). This difference may reflect different actions of local growth factors and is outside the scope of this brief review.

Estrogens modify sexual behavior, control the calcium content of bones in women (decalcification leading to osteoporosis may be inhibited by the action of anti-estrogens) and regulate output of follicle-stimulating hormone and luteinizing hormone by the hypothalamic–pituitary axis.

3.4. Progesterone

The major role of progesterone is to induce differentiation in the endometrial epithelial cells produced by the action of estrogen in the follicular stage of the menstrual cycle. The subsequent appearance of decidual cells indicates that the uterus is ready for implantation of the blastocyst. Progesterone also has a dampening or relaxing effect on the myometrium. Loss of this action of progesterone, at the end of pregnancy, leads to myometrial contraction and parturition. Progesterone is also important in the complete development of the breast and is responsible for terminal end-bud differentiation.

3.5. Androgens

As with estrogens in the female, androgens promote the growth of sexual tissues in the male (prostate and seminal vesicles in the human, comb in the rooster, etc.). Testosterone also stimulates spermatogenesis in the seminiferous tubules. It is responsible for the secondary sexual characteristics (thickening of the vocal chords at puberty, growth of pubic and beard hair, onset of balding, etc.).

Testosterone must first be converted to dihydrotestosterone (DHT) by the enzyme 5α-reductase before it can act in prostate, seminal vesicles, external genitalia, etc. However, in some tissues (kidney, muscles, bone) testosterone is the active hormone.

The biological properties of the steroid hormones are summarized in *Table 1.*

Table 1. Biological actions of steroid hormones

Family name	Common example	Site of synthesis	Target tissues	Actions
Glucocorticoid	Cortisol	Adrenal cortex	Liver, most tissues	Gluconeogenesis Anti-inflammatory Anti-allergic Hypertensive
Mineralocorticoid	Androsterone	Adrenal cortex	Distal tubule of kidney	Na^+/K^+ balance hypertensive
Estrogen	Estradiol-17β	Ovary	Reproductive tissues in female	Endometrial growth Secondary sexual characteristics Bone calcium content, fat deposition at puberty
Progesterone	Progesterone	Ovary	Reproductive tissues in female	Endometrial differentiation Maintains pregnancy Breast differentiation
Androgen	Testosterone	Testis	Reproductive tissues in male	Spermatogenesis Secondary sexual characteristics

4. STEROID HORMONE RECEPTORS

The principal actions of all steroid hormones are mediated by specific steroid receptors. Type I (high-affinity) receptor proteins, with dissociation constants around 10^{-10} M in human target tissues, are found principally within the nucleus of the target cells. The exception is the glucocorticoid receptor, which appears to be evenly distributed in the soluble space of the cytoplasm and nucleus. Lower-affinity receptors (type II receptors) may play a role in storing steroids or helping steroids to cross the cytoplasmic space. Various assays have been developed for detection of steroid receptors by ligand binding, enzyme immunoassay or immunohistochemistry. The practical aspects of these assays have been reviewed (9).

There are differences in the precise details of individual steroid receptors. However, they are all derived from a single superfamily, which has much homology, particularly in the DNA-binding domain (10). The following description applies to the estrogen receptor, but the general concept applies to all steroid receptors. Freshly synthesized receptor monomer is released from its ribosome. The primary sequence contains instructions for forming a dimer and migration to the nucleus. The dimer is a large molecule and passage across the nuclear membrane may be impeded. Since empty estrogen receptor is isolated in conjunction with a dimer of heat shock protein-90 (hsp-90), the concept has evolved that the hsp-90 dimer acts to shepherd the receptor dimer across the nuclear membrane. The receptor dimer then remains in the nucleus but not tightly attached to chromatin. It is argued that the role of the hsp-90 dimer is to bind to and shield the DNA-binding domain of the receptor. Once estradiol ligands arrive in the nucleus, they bind to the complex and induce a rearrangement such that the DNA-binding domain of each receptor molecule becomes exposed, the hsp-90 dimer dissociates (it presumably returns to the cytoplasm) and the estradiol–receptor complex binds tightly to the upstream binding site of estrogen-regulated genes.

The binding of receptor to DNA involves the two zinc fingers in the receptor, which interact with DNA (10). This process is summarized in simplified form in *Figure 2.* It is not proven absolutely that hsp-90 is essential to receptor function in the intact cell. There have been suggestions for roles for both hsp-70 and hsp-27 in receptor function, particularly in relation to the action of anti-steroids.

For gene regulation to take place, the receptor dimer must attract a transcription factor which interfaces between the receptor and RNA polymerase. This acquisition of the transcription factor may be the most important single action of the receptor. Nevertheless, other parts of

Figure 2. A model for the action of estradiol through its specific receptor in a target cell.

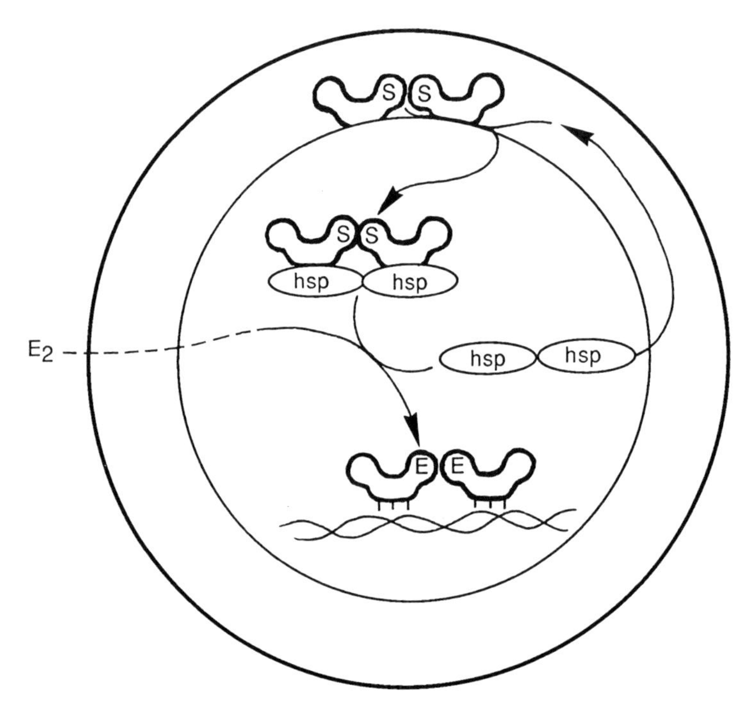

the receptor molecule (e.g. within the steroid-binding domain) do have transactivation effects on gene transcription.

5. SYNTHETIC STEROIDS

Many natural steroids now have synthetic equivalents, which have similar binding affinities for the specific receptor. Four synthetic glucocorticoids have clinical prominence. These are prednisone and prednisolone (both of which bind strongly to CBG) and dexamethasone and betamethasone (both of which have low affinity for CBG). Triamcinolone is also used as a competitor in glucocorticoid-receptor assays. Dexamethasone is commonly used because it has a very high affinity for the glucocorticoid receptor, and so can induce glucocorticoid effects at relatively low doses. Prednisone is used in improving response rates to chemotherapy when used in conjunction with toxic regimes, presumably because it improves the patient's tolerance of the therapy.

Synthetic progestins are used in some contraceptive pills, in some hormone-replacement therapies and in treatment of various hormone-sensitive cancers. There is considerable overlap in their biological activities, which is explained by the fact that megestrol acetate, for example, binds the glucocorticoid receptor almost as well as the progesterone receptor. Because of the high doses given to patients (leading to plasma levels equivalent to 10^{-7} M), it is not always possible to decide whether a response is due to the progesterone receptor or the glucocorticoid receptor.

6. ANTI-HORMONES

Because of the wide variety of actions of steroid hormones, abnormalities in hormone secretion, in availablity of carrier proteins or, more importantly, in availability or function of receptors can lead to serious clinical consequences. For this reason, a whole range of anti-hormones has been developed. Some of these are highly specific. Others interact with several of the steroid receptors. A good example of this is RU486, a molecule that has both anti-progesterone and anti-glucocorticoid activity.

The anti-estrogen, tamoxifen, requires some comment because it is now widely used as adjuvant therapy for almost all patients with breast cancer. Data from the recent overview (11) have shown that all groups of patients (pre- and postmenopausal; node negative and node positive) can benefit from adjuvant tamoxifen, although the extent of benefit is proportional to the amount of estrogen receptor in the primary tumor. At least some of the benefit in receptor-negative tumors appears to come from protection of the heart, as myocardial infarction is much reduced in patients with breast cancer who take tamoxifen compared with equivalent patients who do not (12). Interestingly, tamoxifen has some agonist properties. In studies on the possible use of tamoxifen for the prevention of breast cancer, it has become clear that tamoxifen can stabilize bone mineral content and reduce plasma cholesterol levels (just as effectively as proprietary drugs).

The anti-androgen, cyproterone acetate, also binds to the progesterone receptor which, again, obscures its precise mechanism of action. Flutamide (often in conjunction with the luteinizing-hormone-releasing hormone analogs buserelin or casadex) is now commonly used to block androgen action in prostate cancer.

As can be seen from *Figure 3*, not all anti-steroids are, in fact, steroids, which illustrates the fact that the ability to bind receptor with high affinity is biologically important, rather than having a planar chemical similarity.

Figure 3. The structures of some important anti-steroids.

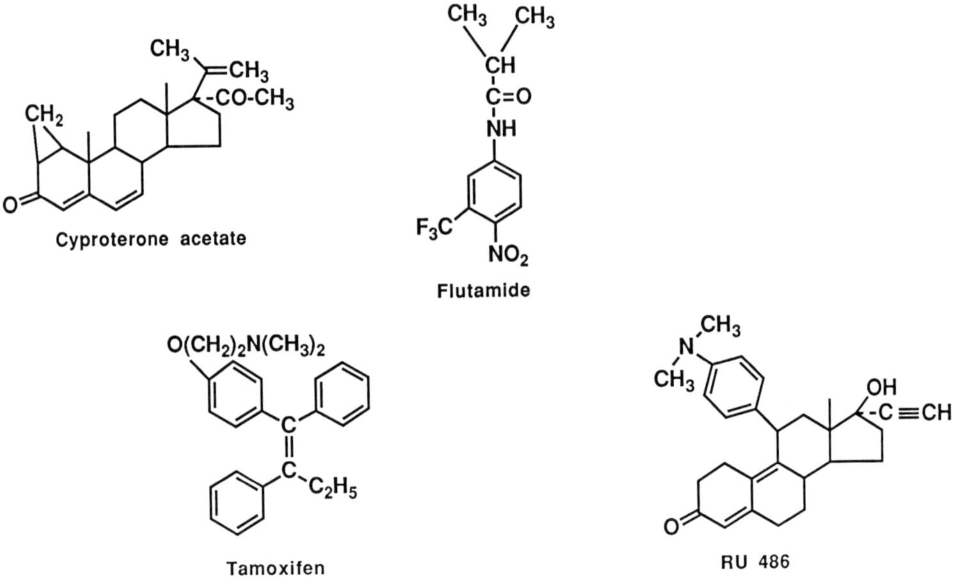

Cyproterone acetate

Flutamide

Tamoxifen

RU 486

In conclusion, the major metabolic interactions of steroids are now well documented. Some of the apparently minor pathways, not mentioned here, may also prove to be of importance. For example, Bradlow and Michnovicz (13) have argued that 16α-hydroxylation of estrogens can lead to breast cancer, whereas promotion of hydroxylation at C-2 may be protective.

Physiological steroid responses are mediated through specific receptors. However, the precise mechanism of activation of receptor, and the interaction of the steroid–receptor complex with individual promoter regions of the genome, are still not fully understood. The pharmaceutical industry is rapidly producing new, synthetic steroids and anti-steroids. These molecules not only throw more light on the biological action of the steroids but also improve the therapy of patients suffering from insufficient or excess steroid activity.

7. REFERENCES

1. Jensen, E.V. and Jacobsen H.I. (1962) *Rec. Prog. Horm. Res.*, **18**, 387.

2. Gower, D.B. (1979) *Steroid Hormones*. Croom Helm, London.

3. Milgrom, E. (1990) in *Hormones — From Molecules to Disease*. (E.-E. Baulieu and P.A. Kelly eds). Chapman and Hall, London.

4. Westphal, U. (1971) *Steroid–protein interactions*. Springer-Verlag, Berlin.

5. Lane, S.J. and Lee T.H. (1991) *Am. Rev. Resp. Dis.*, **143**, 1020.

6. Damon, M., Rabier, M., Loubatiere, J., Blotman, F., and Crastes-de-Paulet, A. (1986) *Agents-Actions.*, **17**, 478.

7. Bulbrook, R.D. and Thomas, B.S. (1989) *Acta Path. Scand.*, **28**, 841.

8. Anderson, T.J., Ferguson, D.J.P. and Raab, G. (1982) *Brit. J. Cancer*, **46**, 376.

9. Leake, R.E. and Habib, F. (1987) in *Steroid Hormones: A Practical Approach*. (B. Green and R.E. Leake eds). IRL Press at Oxford University Press, Oxford.

10. King, R.J.B. (1992) *Clin. Endocrinol*, **36**, 1.

11. Early Breast Cancer Trialists Collaborative Group. (1992) *Lancet*, **339**, 7 and 45.

12. McDonald, C. and Stewart, H.J. (1991) *Brit. Med. J.*, **303**, 435.

13. Bradlow, H.L. and Michnovicz, J.J. (1989) *Proc. Roy. Soc. Edin.*, **95B**, 77.

CHAPTER 12

PEPTIDE HORMONES

D. Patel

1. INTRODUCTION

Peptide hormones, together with the steroid hormones (see Chapter 11) are of central importance for all multicellular animals. These effectors are responsible for the metabolic co-ordination of a very wide range of cells. Peptide hormones can be divided into three major classes based on size, these are large polypeptides (e.g. human chorionic gonadotrophin), oligopeptides (e.g. oxytocin) and amino acid derived hormones (e.g. thyroxine which is derived from a high molecular weight protein). The range of activities of peptide hormones are very diverse, however, some of the most important are those of the pituitary which co-ordinate the functions of a very wide range of cells including the secretion of other hormones.

Many of the peptide hormones act solely at the level of the cell membrane either causing changes in the concentrations of intracellular second messengers or by altering the properties of the membrane directly or indirectly. Hence, the activities of these hormones in modifying the transport properties of membranes are described in Chapter 8 of this book. This chapter will restrict itself to listing the key features of peptide hormones in terms of their structure function and target cells.

2. ABBREVIATIONS

♂, males; ♀, females; cAMP, cyclic 3′,5′ adenosine monophosphate; CNS, central nervous system; conc., concentration; Inhibs and (I), inhibitors; insol., insoluble; mol. wt, molecular weight; NADPH, reduced nicotinamide adenine dinucleotide phosphate; no., number; P_i, inorganic phosphate; sl. sol., slightly soluble; sol., soluble; Stims and (S), stimulators; temp., temperature.

Hormone abbreviations: ACTH, adrenocorticotropic hormone; ADH, antidiuretic hormone (Vasopressin); CRH, corticotropin-releasing hormone; FRH, follicle-stimulating hormone-releasing hormone; FSH, follicle-stimulating hormone; GH, growth hormone; GRH, growth hormone-releasing hormone; ICSH, interstitial cell stimulating hormone, in men; LH, luteinizing hormone (interstitial cell stimulating hormone), in women; LH-RH, luteinizing hormone-releasing hormone; LTH, lactogenic hormone; MRIH, melanocyte-stimulating hormone release-inhibiting hormone; MRH, melanocyte-stimulating hormone release hormone; MSH, melanocyte-stimulating hormone; PRIH, prolactin release-inhibiting hormone; PTH, parathyroid hormone; TRH, thyrotropin-releasing hormone; TSH, thyrotropic hormone.

Amino acid abbreviations: Acetylseryl, AcSer; Alanine, A; Arginine, R; Asparagine, N; Aspartic acid, D; Cysteine, C; Diidotyrosyl, diido Tyr; Glutamine, Q; Glutamic acid, E; Glycine, G; Histidine, H; 3-Iodotyrosyl or 3-iodotyrosine, 3-iodo Tyr; Isoleucine, I; Leucine, L: Lysine, K; Methionine, M; Phenylalanine, F; Proline, P; Q-Pyroglutamyl, (pyro) Q; Serine, S; Threonine, T; Tryptophan, W; Tyrosine, Y; Valine, V; Unknown or 'other', X.

Table 1. Peptide and protein hormones

Hormone (synonym)	Structure and properties	Sources	Targets	Principle functions and effects	Inhibs(I) and Stims(S) of secretion	Reference
Corticotropin-releasing hormone (Adrenocorticotropin-releasing hormone; CRH or factor: CRF; at least two structures, α & β)	Basic peptide, with NH_2-terminal structure like that of α-MSH; and COOH-terminal structure like that of lysine vasopressin; tentative structure of β-CRH : AcSer-Y-C-F-H-N-E-P-V-K-G-NH_2; inactivated by oxidation, reduction, peptic and tryptic proteolysis, disulfide linkages	Hypothalamus and neurohypophysis of man, cattle, dog, rabbit, rat, sheep and swine	ACTH-secreting cells of adenohypophysis	Stimulates: release of ACTH from basophils of adenohypophysis into blood; also increases rate of synthesis of ACTH by basophilic cells	(I): High blood levels of corticosteroids (long-term) and of ACTH (short-term) — negative feedback from suprarenal cortex normally controls secretion of CRH; final target organ probably more important (S): Neural impulses from CNS; neurosecretory cells function as transducers; low blood levels of corticosteroids and ACTH; cyclic nyctohemeral control through hypothalamus; stress	1
Follicle-stimulating-hormone-releasing hormone (Follicle-stimulating-hormone-releasing factor; FRH; FSH-RH; FSH-RF)	Amino acid sequence: (pyro) Q-H-W-S-Y-G-L-R-P-G-NH_2 sol. in 0.1 M HCl; stable to heat	Ventral hypothalamus of man, cattle, rat, sheep and swine	FSH-LH-secreting basophils (δ-cells) of adenohypophysis	Stimulates: release of FSH and LH into peripheral blood; ovulation	(I): High levels of FSH or of estrogens in blood (negative feedback) (S): Neural stimulation from CNS; constant exposure to light; low levels of FSH or of estrogens; dopamine	1, 2

Table 1. Continued

Hormone (synonym)	Structure and properties	Sources	Targets	Principle functions and effects	Inhibs(I) and Stims(S) of secretion	Reference
Growth-hormone-releasing hormone (Growth-hormone-releasing factor; somatotropin-releasing hormone; GRH, GRF; SRH; SRF)	A polypeptide; acidic; mol.wt = 2500 Da; stable in boiling 0.1 M HCl; inactivated by proteolysis, not inactivated by thioglycollate	Hypothalamus of man, cat, cattle, dog, guinea-pig, rat and swine	GH-synthesizing acidophils (alpha-cells) of adenohypophysis	Stimulates: release of GH from adenohypophysis into blood	(I): High blood levels of GH (negative feedback) (S): CNS stimulation; decreased blood levels of GH (depletion of GH in the hypophysis)	1, 2
Luteinizing-hormone-releasing hormone (Luteinizing-hormone-releasing factor: LH-RH; LRH; LH-RF; LRF)	Amino acid sequence is same as for FRH; stable in boiling 0.1 M HCl	Hypothalamus of man, cattle, dog, monkey, rabbit, rat, sheep and swine	ICSH-(LH-) secreting basophils of adenohypophysis	Stimulates: release of ICSH(LH) into blood Decreases: ovarian ascorbic acid	(I): High levels of LH and of estrogens or progesterone in blood (negative feedback); suckling stimulus. (S): Neural stimulation from CNS; proestrus, puberty; exposure to light	1, 2
Thyrotropin-releasing hormone (Thyrotropin-releasing factor; thyroid-stimulating-hormone-releasing hormone; TRH; TRF)	Bovine, ovine, porcine sources: L-(pyro) Q-H-P-NH_2 (synthetic tripeptide has activity); sol. in dilute acetic acid; destroyed by boiling in 6 M HCl; not inactivated by proteolysis	Hypothalamus and neurohypophysis of man, cattle, goat, rabbit, rat, sheep and swine	TSH-secreting cells of adenohypophysis	Stimulates: release of TSH from basophils of adenohypophysis: secretory phase of thyroid gland	(I): High blood levels of TSH and thyroid hormones (negative feedback) (S): Neural impulses from CNS; low blood levels of thyroid hormones and TSH	1,2

Table 1. Continued

Hormone (synonym)	Structure and properties	Sources	Targets	Principle functions and effects	Inhibs(I) and Stims(S) of secretion	Reference
Prolactin release-inhibiting hormone (PRIH; PRH; PIF)	Polypeptide of low mol wt.; different from epinephrine, oxytocin and ADH or any kinin; sol. in 0.1 M HCl; stable to boiling	Same part of hypothalamus that secretes ICSH of cattle, rat, sheep and swine	LTH-secreting basophils of adenohypophysis	Inhibits: release of LTH from adenohypophysis of ♂ and ♀	(I): Lactation; high blood levels of PRIH (negative feedback); increased blood levels of gonadal steroids (S): Unknown; presumably excessive lactation	1
Natriuretic hormone	Polypeptide; one basic amino acid, one free terminal NH_2 group; mol. wt <20 kDa; dialyzable; stable to acid, heat; inactivated by proteolysis by aminopeptidase, chymotrypsin and trypsin	Hypothalamus; posterior nucleus; plasma	Epithelial cells in proximal tubule of kidney	Inhibits (regulates?): Na^+ transport (reabsorption) from lumen of proximal tubule of kidney into capillary — much more effective than oxytocin or ADH	(I): Unknown (S): Released from posterior hypothalamus by hemodynamic stimuli (e.g. carotid occlusion)	3
Somatostatin (SRIF; Somatotropin [growth hormone] release-inhibiting factor)	Polypeptide; mol.wt = 1638 Da; amino acid sequence: NH_2-A-G-C-K-N-F-F-W-K-T-F-T-S-C; both oxidized (disulfide) and reduced (dithiol) forms have biological activity		CNS outside the hypothalamus and intestinal tract	Inhibits release of somatotropin by anterior pituitary, and of thyrotropin (but not of prolactin); inhibits secretion of insulin, glucagon and gastrin		4
Adrenocorticotropic hormone (Adrenocorticotropin; corticotropin; ACTH)	Unbranched polypeptide chain of 39 amino acids; NH_2-terminal Ser, COOH-terminal Phe; residues 1–24 essential for activity in all species; mol. wt = 4500 ± 50 Da (varies	Adenohypophysis-chromophobic basophils of pars distalis	Zonae fasciculata and reticularis of suprarenal cortex; mitochondria	Decreases: ascorbic acid and cholesterol in suprarenal cortex; renal transport of P_i and urate; urea production from exogenous amino acids; liver degeneration of corticosteroids	(I): Increased blood levels of glucocorticoids (negative feedback, or servomechanism) — mediates neural regulation of	2, 5

Table 1. Continued

Hormone (synonym)	Structure and properties	Sources	Targets	Principle functions and effects	Inhibs(I) and Stims(S) of secretion	Reference
	among species); pI = 4.7–4.8; sol. in water, stable in hot acids; destroyed by 0.1 M NaOH at 100°C		and enzyme systems in cytoplasm of many tissues; chromatophores	Increases: secretion of cortical hormones; oxidative phosphorylation; lipolysis in adipose tissue; fatty acid transport and oxidation; protein synthesis; iodine uptake by thyroid; synthesis of cAMP, activating α-glucan phosphorylase and increasing NADPH in suprarenal cortex; stimulates DNA synthesis by activating DNA nucleotidyl-transferase	corticoid secretion; blocked by dexamethasone (S): CRH of hypothalamus or neurohypophysis; stimulation of median eminence; psychic trauma, acting through hypothalamus; decreased level of circulating glucocorticoids, mainly hydrocortisone	
Follicle-stimulating hormone (FSH)	Glycoprotein containing sialic acid and small amounts of hexoses and hexosamines (up to 20% carbohydrate in man); mol.wt = 36 kDa for man (dimer); sol. in water, half-saturated $(NH_4)_2SO_4$; stable at pH 7–8 for 30 min at 75°C; destroyed by reducing disulfide bonds	Basophils of adenohypophysis of man (♂♀), sheep (♂♀), and swine (♂♀); post-menopausal serum and urine	Seminiferous tubules (♂); graafian follicles of ovary (♀)	Stimulates: growth and maturation of ovarian follicles Increases: spermatogenesis; growth of seminiferous tubules; testosterone secretion; supporting (Sertoli) cell hormone secretion in testis; incorporation of α-amino-isobutyric acid into proteins of ovaries (rat); growth and maturation of ovarian follicles — small amounts of LH required; transport of glucose and amino acids across cell membrane in ovary	(I): Increased levels of circulating estrogens (S): Low blood levels of estrogens or possibly androgens; castration; menopause; hypothalamic–neurohypophyseal stimulation; external factors (e.g. light) through hypothalamus; FRH of hypothalamus	2, 5

Table 1. Continued

Hormone (synonym)	Structure and properties	Sources	Targets	Principle functions and effects	Inhibs(I) and Stims(S) of secretion	Reference
Growth hormone (Somatotropin; GH)	Polypeptide (size varies with species); NH_2-terminal and COOH-terminal amino acid is Phe for all species, but also Ala for NH_2-end in cattle and sheep; no. of disulfide bonds in GH in man = 2; mol.wt = 21.7 kDa in man; pI = 5.12 in man; sol. in dilute solutions of neutral salts and in ethanol; sl. sol. in water; inactivated by HNO_2, acetylation	Acidophils of adenohypophysis; plasma (in man, GH levels decrease with age from 2 to 16 yr)	Granular endoplasmic reticulum of liver and most tissues; epiphyseal cartilage; fibroblasts	Decreases: blood amino acid and glucose levels after short-term administration; urinary excretion of inorganic ions Increases: skeletal and soft-tissue growth; protein anabolism; fibroblastic activity; swelling of liver mitochondria (rat); activity or synthesis of RNA nucleotidyl-transferase (RNA polymerase); transport of neutral amino acids, glucose and other hexoses; transport of fatty acids from fat depots to liver, with lowering of the respiratory gas-exchange ratio; produces positive balances of Na^+, K^+, N_2 and P_i into tissues; erythropoiesis; renal function; tubular reabsorption of SO_4^{2-} (dog); blood glucose levels after prolonged administration	(I): Hyperglycemia or high glucose-glycogen metabolism; maintenance of positive nitrogen balance; increased plasma corticosteroid levels (Cushing's syndrome); hyperlipemia; increase in plasma levels of GH; not species specific (S): Hypoglycemia, either fasting- or insulin-induced; exercise while fasting; prolonged fasting; emotional or traumatic stress; rapid growth (i.e. protein synthesis in infants); pregnancy (?); anorexia nervosa; intravenous infusion of Arg; hypolipemia; GRH of hypothalamus; L-dopamine causes rise in human GH persisting < 2 h; increased secretion of GRH by neural stimulation	2, 5, 6, 7, 8

Table 1. Continued

Hormone (synonym)	Structure and properties	Sources	Targets	Principle functions and effects	Inhibs(I) and Stims(S) of secretion	Reference
Interstitial cell-stimulating hormone (ICSH, for ♂; Luteinizing hormone, LH, for ♀)	Glycoprotein containing hexose, sialic acid, galactosamine and glucosamine; adsorbed on cation-exchange resins; mol.wt = 33.15 kDa in man (dimer); pI = 5.4 for man; is sol. in water, dilute solutions of acidic or neutral salts, 40% ethanol; in man up to 5% carbohydrate content (ICSH); 18% carbohydrate (LH)	Basophils of adenohypophysis; postmenopausal plasma or urine	In ♂, interstitial (Leydig) cells of testes; in ♀, graafian follicles 'primed' to maturity by FSH and corpora lutea of ovaries	Increases (♂); spermiogenesis by seminiferous tubules; biosynthesis of testosterone by interstitial (Leydig) cells — synergistic with FSH Increases (♀): follicle maturation, with FSH; production of corpora lutea — requires FSH and LTH for maintenance; estrogen and progesterone biosynthesis by corpora lutea; uptake of glucose by ovaries	(I): High blood conc. of progesterone or testosterone, fluctuating with menstrual cycle in ♀; plasma levels much higher in postmenopausal ♀ (S): Low or moderate blood levels of gonadal hormones; LH–RH of hypothalamus	2, 5, 9, 10, 11
Lactogenic hormone (Luteotropin; prolactin; mammotropin; LTH, MH)	One peptide chain; 205 amino acid residues in sheep; in cattle and sheep no. of disulfide bridges is six; in man, lactogenic hormone similar to but distinct from human growth hormone; mol. wt = 23 kDa in sheep; pI = 5.73 in cattle and sheep; sol. in acid methanol, dilute acids; sl. sol. in water, dilute salt solution; not inactivated by partial hydrolysis	Basophils of adenohypophysis of cattle, rat, sheep and swine	Corpora lutea, matured by LH; secretory cells of mammary glands; pigeon crop	Increases: milk secretion ejection by mammary glands prepared by estrogen and progesterone; nidation of zygote; protein anabolism (GH-like in most species) in mammals, lipogenesis and fat deposition (also in birds and teleosts), sebaceous gland size and activity, and growth of ♂ accessory sex organs (synergistic with androgens)	(I): Inhibition of oxytocin release; nervous inhibition; PRIH from hypothalamus (S): Oxytocin, in turn, stimulated by suckling	1, 5, 12

Table 1. Continued

Hormone (synonym)	Structure and properties	Sources	Targets	Principle functions and effects	Inhibs(I) and Stims(S) of secretion	Reference
Thyrotropic hormone (Thyrotropin; TSH)	Glycoprotein (single peptide chain) containing covalently bound fucose, mannose, galactosamine and glucosamine (species differences, but immunologic cross-reaction among TSH from man, cattle and swine); COOH-terminal amino acid sequence in man and cattle = H-V-K-S-V; mol. wt = ~ 28 kDa in man; inactivated by boiling, Cys ketene; similar in structure and action to LH; up to 13% carbohydrate content in man	Basophils of adeno-hypophysis; blood	Secretory epithelial cells of thyroid gland; adipose tissue cells(?)	Decreases: iodine and colloid content of thyroid Increase: synthesis and mobilization of thyroid hormones; serum protein-bound I; thyroid RNA and protein; proteolytic activity; oxidase granules; O_2 consumption by thyroid cells; entry of iodide into thyroid cells; entry of glucose into cells; rate of glucose oxidation by hexose monophosphate, glycolytic and tricarboxylic acid pathways; lypolysis in adipose tissue; activities of enzymes mediated by cAMP, e.g. iodine-trapping and 'organification', coupling of mono- and diido(Tyr), NADPH oxidation to pentose shunt, uptake of glucose and amino acids, phospholipid turnover	(I): Increased circulating thyroid hormones; high ambient temp; inhibition of hypothalamus (S) Decreased circulating thyroid hormones; low ambient temp.; TRH of hypothalamus	1, 2, 5, 9, 13, 14
Melanocyte-stimulating hormone (Melanotropin; MSH; melano-phore-stimulating hormone)	Dialyzable; smallest sequence retaining activity is M-Q-H-F-R-W-G; adsorbed on oxycellulose; sol. in water, acidified acetone; moderately heat-stable; destroyed by tryptic proteolysis	Pars intermedia of hypo-physis of many species including mammals; anterior lobe in birds and	Melanocytes of skin — most noticeable in species with chromato-phores, but active in all species	Expands chromatophores rapidly (within a few min), causing pigment granules to disperse and color skin *in vivo* or *in vitro* (not species-specific — MSH from animals expands chromatophores in human	(I): High blood level of corticosteroids(?); MRIH from hypothalamus (S): Low level of corticosteroids (e.g. Addison's disease); MRH of	2, 5

Table 1. Continued

Hormone (synonym)	Structure and properties	Sources	Targets	Principle functions and effects	Inhibs(I) and Stims(S) of secretion	Reference
		porpoises; synthesized *in vitro*		skin); increases during most of pregnancy; expands chromatophores, contracts guanophores in anuran larvae	hypothalamus; sympathetic postganglionic potentials from superior cervical ganglion	
α-MSH	Tridecapeptide; same residues as those for ACTH; pI = 10.5–11; amino acid sequence: AcSer-Y-S-M-Q-H-F-R-W-G-K-P-V-NH_2	Basophils of hyprophysis of all mammals; synthesized *in vitro*		Increases: synthesis of melanin by chromatophores in human skin — most apparent in areas under pressure or trauma (scars)		
β-MSH	18 amino acid residues in cattle, sheep and swine, 22 in man; core sequence essential for activity in all species is M-Q-H-F-R-W-G; pI = 4.1	Basophils of hypophysis; synthesized *in vitro*				
Oxytocin (Oxytocic hormone)	Octapeptide: C-Y-I-E-N-C-P-L-G-NH_2; inactivated by reduction of disulfide or by alkali; hydrolyzed by oxytocinase in serum of pregnant women, pepsin and chymotrypsin	Neurohypophysis; blood of man, cattle, horse, rat and sheep; synthesized *in vitro*	Uterine and other smooth muscle; myoepithelial cells of mammary alveoli	Facilitates: sperm movement up fallopian tubes Stimulates: release of LTH Increases: uterine muscle contraction — may initiate labor; permeability of myometrial cell membrane to K^+, thus decreasing membrane potential and excitability threshold	(I): No specific inhibitors known (may be a feedback mechanism) (S): Stimuli from birth canal; in animals with multiple fetuses, suckling of firstborn stimulates birth of more young (neural through hypothalamus); oxytocin level rises sharply half-way through labor	1, 15

Table 1. Continued

Hormone (synonym)	Structure and properties	Sources	Targets	Principle functions and effects	Inhibs(I) and Stims(S) of secretion	Reference
Vasopressin (Antidiuretic hormone; ADH)	Octapeptide: C-Y-F-E-N-C-P-R-G-NH_2; disulfide bond essential for activity; mol. wt = 1228; pI = 10.8; sol. in acid ethanol, phenol; stable in acid, not in alkali; inactivated by vasopressinase in serum of ♂ and nonpregnant ♀, by oxytocinase in the serum of pregnant ♀ and by trypsin and chymotrypsin, but not by pepsin or pancreatic carboxypeptidase	Neurohypophysis and blood of man, cattle, dog, horse, rat and sheep	Capillaries; arterioles; coronary vessels; vascular bed and tubules of kidney; smooth muscle	Increases: water reabsorption in renal tubules by increased membrane permeability in distal tubules; excretion of Na^+, Cl^- and urea; blood pressure by arteriole constriction; activity of adenyl cyclase, and thereby level of cAMP in blood and sizes of water 'pores' in distal and collecting tubules (possibly forms a disulfide bond with cell membrane)	(I): Decreased blood plasma osmotic pressure; increased extracellular fluid volume (S): Increased blood osmotic pressure, stimulating osmoreceptors in diencephalon; decreased volume of extracellular fluid (e.g. hemorrhage), stimulating volume ('stretch') receptors; nephrogenic stimuli through hypothalamus; drugs, e.g. morphine, nicotine, anesthetics	2

Table 1. Continued

Hormone (synonym)	Structure and properties	Sources	Targets	Principle functions and effects	Inhibs(I) and Stims(S) of secretion	Reference
Thyroglobulin	Glycoprotein containing A, R, D, C, Q, G, H, I, L, K, M, F, P, S, T, W, Y, V, diiodo(Y), T_4 and iodo-T_4; mol.wt = 660 kDa; pI = 4.6; in man up to 11% carbohydrate content	Follicular colloid of thyroid gland (stored form of thyroid hormones); hydrolyzed by proteases in the colloid	Most cells of the body, especially bones, striated muscle, heart, liver and kidney — synergistic with GH	In large amounts, uncouples phosphorylation from oxidation in mitochondria Stimulates: growth; maturation; neuromuscular function; skin development; hematopoiesis; gametogenesis; lactation; O_2 consumption; absorption through intestinal wall Decreases: TSH secretion Increases: rate of O_2 uptake without increased P_i uptake; rate of protein synthesis in cells by increasing RNA synthesis; rate of lipid, protein, carbohydrate, mineral and water metabolism	(I): Severe stress, such as fear, rage, hemorrhage, inflammatory processes and severe muscular exercise inhibits secretion for first 24–48 h; prolonged stress stimulates increased uptake of I^- or of antithyroid drugs, e.g. 2-thiouracil, thiourea, thiocyanate; increased levels of thyroid hormones; high ambient temperature (S): Low ambient temp.; pregnancy; prolonged stress; TSH; prolonged dietary I^- deficiency; long-acting thyroid stimulator (LATS) in plasma of thyrotoxic patients (an A 7S γ-globulin, possibly an immuno-globulin)	1, 2, 5, 16

Table 1. Continued

Hormone (synonym)	Structure and properties	Sources	Targets	Principle functions and effects	Inhibs(I) and Stims(S) of secretion	Reference
Thyrocalcitonin (Calcitonin; TCT)	Polypeptide; 32 amino acid residues; a 1,7-disulfide ring at NH_2-terminal end; crude preparations stable at neutral pH and room temp. (purified preparations labile); crude and purified preparations stable at pH 3; no loss of activity by alkylation or oxidation of Met; inactivated by reduction of disulfide bond, proteolysis by trypsin; chymotrypsin or pepsin, and oxidation by *o*-diphenol oxidase (pyrocatechol oxidase), H_2O_2 or light	Tissue of ultimo-branchial origin (C cells, parafolli-cular cells, 'light' cells) in thyroid; in man, in parathyroids and thymus; distinct ultimo-branchial bodies are main source in birds, reptiles and fish	Blood–bone membrane of osteoblasts, osteoclasts and osteocytes	Causes hypophosphatemia Regulates: Ca^{2+} homokinesis; skeletal development and maintenance by altering Ca^{2+} exchange Inhibits: bone resorption, by decreasing permeability of bone membrane to Ca^{2+}, thereby producing hypocalciuria and possible hypocalcemia, depending on dietary intake of Ca^{2+} and vit. D; requires vit. D for action	(I): Decreased serum Ca^{2+} in pathological hypocalcemias (S): Increase of Ca^{2+} above 10 mg/100 ml	1, 2, 5
Parathyroid hormone (Parathormone; PTH)	Polypeptide; 84 amino acid residues in man, cattle and swine; first 29 residues have been synthesized and are required for activity; 20 amino acids at COOH-terminal end are active portion of molecule; mol. wt= ~8500 Da (varies with species); is sol. in water, saline, aqueous ethanol, 94% acetic acid, conc.	Parathyroid glands; malignant tumors, especially carcinomas of lung and kidney; urine, especially in hyperpara-thyroid humans	Bone–blood membrane of osteo-blasts, osteoclasts and osteo-cytes; cells of renal proximal tubules and of intestinal mucosa(?)	Regulates: Ca^{2+} and P_i levels in serum by mineral exchange between blood and bone, and by Ca^{2+} reabsorption and P_i excretion by kidney; renal tubule excretion of K^+, HPO_4^{2-} and $H_2PO_4^-$ (actions on both bone and renal cells mediated through increased activity of cAMP; increase of PTH	(I): Increased conc. of free Ca^{2+} (from diet or bone); decreased conc. of P_i(?). (S): Decrease of serum Ca^{2+} below 10 mg/100 ml; increased serum P_i	1, 2, 5

Table 1. Continued

Hormone (synonym)	Structure and properties	Sources	Targets	Principle functions and effects	Inhibs(I) and Stims(S) of secretion	Reference
	phenol, 50% glycerol; is insol. in volatile organic solvents; stable in dilute HCl; inactivated by oxidation of Met, Trp or Tyr, or by acetylation or esterification			causes increased cAMP in both tissues; vit. D required) Stimulates: DNA and RNA synthesis and binding of enzymes, cAMP and Ca^{2+} to membranes of bone cell		
Glucagon	Single polypeptide chain of 29 amino acids; mol. wt = 3485 Da; pI = 7.5–8.5; sol. in dilute acid and alkali; insol. in water; stable in alkali; forms fibrils when heated to 40°C in acidic solution and then cooled	α cells of islets of Langerhans in pancreas	Similar to insulin	Antagonistic to insulin Inhibits: synthesis of fatty acids from precursors Promotes: glycogenesis in liver but not in muscle; glycogenolysis in liver by stimulation of cAMP synthesis which mediates: (i) conversion of phosphorylase *b* to α-glucan phosphorylase (phosphorylase *a*); (ii) release of non-esterified fatty acids from adipose tissue by glucagon and epinephrine (iii) increased protein metabolism; (iv) stimulatory effect of epinephrine on cerebral cortex; and (v) increased force of heartbeat (inotropic effect) by glucagon and epinephrine	(I): Hyperglycemia; glucagon and insulin regulate blood levels by opposite feedback mechanisms, involving liver glycogen as storage compound; increased conc. of plasma fatty acids (S): Hypoglycemia; pancreozymin — potency in stimulating glucagon secretion 1/6 that of insulin secretion; increased conc. of plasma amino acids	2, 5, 6

Table 1. Continued

Hormone (synonym)	Structure and properties	Sources	Targets	Principle functions and effects	Inhibs(I) and Stims(S) of secretion	Reference
Insulin	Polypeptide consisting of an A chain (21 amino acids) and a B chain (30 amino acids) connected by disulfide bridges; species differences in residues 8, 9 and 10 of disulfide ring in A chain and in COOH-terminal residue of B chain (no effect on activity); more than one type may be present in a single species; crystallized with Zn^{2+} (3 atoms/mol.wt of 35 kDa); mol.wt = 5734 Da at pH 2.5 in cattle (minimum), possibly 12–48 kDa at pH 7–8; unknown in circulating blood; pI = 5.3; stable in acid solution; resistant to denaturing agents; inactivated in alkaline solution; completely inactivated by cleavage of disulfide bonds between A and B chains by oxidation or reduction, and by proteolysis	β cells in islets of Langerhans in pancreas; proinsulin (an inert peptide chain of 33 residues connecting A and B chains), synthesized in granular endoplasmic reticulum, transferred to Golgi bodies, and packaged in granules; hydrolyzed to insulin (sol. insulin also present); released into circulation by appropriate stimulation; synthesized *in vitro*	All tissues and cells involved in metabolism of carbohydrates, lipids and proteins, enzymatic systems in cell membranes, mitochondria and other cytoplasmic organelles of muscle, adipose tissue and liver	Antagonistic to glucocorticoids Regulates: carbohydrate and fatty acid metabolism Activates: ribosomes; glycogen synthetase Stimulates: glycolytic enzymes; phosphorylases; lipogenesis; transport and oxidation of lipids; oxidation of carbohydrates; glucose and amino acid uptake by cells; synthesis of protein and mucopolysaccharides Decreases: gluconeogenesis; ketogenesis; lipolysis in adipose tissue; free sulfhydryl content of muscle; activity of adenylyl cyclase, and thereby the conc. of cAMP Increases: rate of transport of glucose into striated muscle and adipose tissue, and of K^+ into muscle; synthesis of mRNA (30 min) and DNA (24–36 h) in liver, stimulating protein synthesis upon increased uptake of amino acids by striated muscle, coupled with transport of Na^+; activities of lipogenic enzyme systems, and	(I): Hypoglycemia during fasting or prolonged exercise; hyperinsulinemia (negative feedback ?); epinephrine; norepinephrine — α-adrenergic stimulation (S): Hyperglycemia (glucose must be phosphorylated or further metabolized); high-protein diet (increased amino acids in blood); stimulants of adenylyl cyclase synthesis; short-chain fatty acids; sulfonylureas; tolbutamide; cAMP; gastrointestinal hormones — gastrin, pancreozymin and secretin; glucagon?; increased ACTH, GH; vagal stimulation; all stimulants probably induce	2, 5, 6

Table 1. Continued

Hormone (synonym)	Structure and properties	Sources	Targets	Principle functions and effects	Inhibs(I) and Stims(S) of secretion	Reference
				thereby the deposition of triglycerides in adipose tissue; incorporation of SO_4^{2-} chondroitin sulfates	synthesis of proinsulin by β cells and facilitate release of insulin into blood	
Cholecystokinin-pancreozymin (CCK-PZ)	Polypeptide consisting of 33 amino acid residues (low mol. wt, dialyzable); identical to pancreozymin; pI = 5.0–5.5; is sol. in water	Mucosa of upper intestine	Muscle of gallbladder (may be identical with pancreozymin)	Increases: contraction and emptying of gallbladder (cholagogue effect)	(I): Anticholecystokinin in blood and urine secreted by wall of gallbladder (S): Fatty, fatty acids, polypeptides or acidic chyme in duodenum	1, 2
Gastrin	Polypeptide; active portion pentapeptide G-W-M-D-F-NH_2; dialyzable; pI = 5.5; stable to heat; destroyed by u.v. light, alkali, peptic hydrolysis	Gastric mucosa, mainly in pyloric antral area, of man, dog, sheep and swine; synthesized *in vitro*	Parietal and chief cells of gastric mucosa: biliary system of liver	Stimulates: secretion of gastric HCl, pepsinogen and Castle's intrinsic factor; also liver bile and pancreatic enzymes	(I): High H^+ conc. in stomach (S): Polypeptides in contact with pyloric mucosa	1, 17
Pancreozymin-Cholecystokinin (PZ-CCK)	Polypeptide; identical to cholecystokinin; physical and chemical properties very similar to those of cholecystokinin; stable to heat, acid; destroyed by alkali	Mucosa of upper intestine (first meter)	Enzyme-secreting cells of pancreas (may be identical with cholecystokinin)	Inhibits: gastric motility and secretion Stimulates: secretion of enzymes by pancreas, possibly by promoting transport across cell membranes (no effect on volume)	(I): Absence of duodenal chyme (S): Fatty acids; fats; polypeptides; certain amino acids	1, 17

Table 1. Continued

Hormone (synonym)	Structure and properties	Sources	Targets	Principle functions and effects	Inhibs(I) and Stims(S) of secretion	Reference
Secretin	Polypeptide, with 27 amino acid residues; strongly basic; synthetic polypeptide has full activity; sol. in dilute acids, alcohols; salted out by NaCl, trichloroacetic acid	Mucosa of upper intestine	Acinar or exocrine cells of pancreas; biliary system of liver	Inhibits: gastric motility and secretion Stimulates: volume flow rate of pancreatic juice, including ions but not enzymes; bile flow from liver; intestinal juices	(I): Absence of food in duodenum (S): Acidic chyme, water, ethanol, fatty acids, polypeptides, some amino acids in duodenum	1, 17
Human chorionic gonadotrophin (HCG)	Glycoprotein containing sialic acid, hexoses, hexosamines and amino acids (Y-R-W), 135 000 IU mg^{-1}; mol. wt = 55–60 kDa (dimer?); pI = 2.95; sol. in water, 50% acetone, 60% ethanol; inactivated by removal of *N*-acetylneuraminic acid (NANA) end-groups; up to 35% carbohydrate content in man	Chorionic villi of human placenta; blood and urine of pregnant women — peak at 8th week of pregnancy (60–75 days after 1st day of last menses), decreasing to 15% of peak at week 16	Ovaries, especially corpora lutea	Actions similar to those of LH and LTH: synergistic with hypophyseal gonadotropins; maintains corpora lutea during early pregnancy Stimulates: luteal production of progesterone until placenta attains full function; premature descent of testes in immature primates Increases (by injection); follicle maturation and ovum release in nonpregnant ♀; androgen secretion by interstitial (Leydig) cells of testes	(I): ♀ sex steroids, notably progesterone (S): Enzyme systems of developing placenta (specific stimulant unknown)	1, 2, 5, 18
Human placental lactogen (HPL)	Mol.wt = 30–38 kDa (dimer, or two chains?)	Villus trophoblast of human placenta; blood of pregnant women (conc. higher than that of human GH)	Myoepithelial cells of mammary glands; corpora lutea	Synergistic with progesterone, LTH and oxytocin in stimulating and maintaining lactation, and in milk ejection; augments GH	(I): Unknown (S): Not definitely known; probably gonadotropins or GH	1

Table 1. Continued

Hormone (synonym)	Structure and properties	Sources	Targets	Principle functions and effects	Inhibs(I) and Stims(S) of secretion	Reference
Relaxin	Polypeptide; mol.wt = 9 kDa; pI = 5.5; is sol. in water, 95% ethanol	Corpora lutea of pregnant women, mouse, swine; placenta; endometrium; blood of pregnant women, cat, dog, horse, rabbit and swine	Pubic ligaments	Effects relaxation of pubic ligament and separation of symphysis pubis, chiefly during last stages of pregnancy in most mammals; after pre-sensitization with estrogen, connective tissues of symphysis become more vascular, collagen fibers are dissolved, and mucopolysaccharides are depolymerized, rendering ligament more flexible; effects softening and relaxation of uterine cervix after pretreatment with estrogen; water imbibition by myometrium in rat; synergistic with estrogen and progesterone in mammary development in rat Inhibits: rhythmic uterine contractions (in rat) and uterine motility	(I): Not definitely known; possibly changes in balance of estrogens and progesterone during parturition (levels in blood peak during terminal stages of pregnancy and decline rapidly just prior to, during, and 1 day after delivery) (S): Not definitely known; probably increased levels of progesterone during pregnancy	17, 19–21

Table 1. Continued

Hormone (synonym)	Structure and properties	Sources	Targets	Principle functions and effects	Inhibs(I) and Stims(S) of secretion	Reference
Angiotensin (Hypertension; Angiotonin)	Angiotensin-II is the more active form; angiotensin-I has much lower activity; angiotensin-I is formed by the action of renin on angiotensinogen in the plasma, and is converted into angiotensin-II by converting enzyme, which removes the C-terminal dipeptide; angiotensin-II is short-lived in blood and in tissues due to further degradation by peptidases; stable in neutral solution; hydrolyzed by strong acids and above pH 9.5; is sol. in water, aqueous solutions pH 5–8 and ethanol; in man, amino acid sequence: angiotensin-I: D-R-V-Y-I-H-P-F-H-L, mol. wt = 1297 Da; angiotensin II: D-R-V-Y-I-H-P-F, mol. wt = 1049 Da	Plasma	Skeletal, mesenteric and renal vascular beds; adrenal gland	Raise blood pressure (pressor action) by causing vasoconstriction in skeletal, mesenteric and renal beds; stimulates release of aldosterone from adrenal gland		4
β-Endorphin	Naturally occurring opioid peptide; mol.wt = 3464 Da; β-endorphin has same amino acid sequence as residues 61–91 of β-lipotropin		Opiate receptor sites	Possesses potent agonist activity at opiate receptor sites	Action antagonized by naloxone	4

Table 1. Continued

Hormone (synonym)	Structure and properties	Sources	Targets	Principle functions and effects	Inhibs(I) and Stims(S) of secretion	Reference
Enkephalin	Naturally occurring opioid peptide; unstable in blood; Tyr is cleaved from N-terminus; $[Met^5]$-Enkephalin has mol. wt = 574 Da and amino acid sequence Y-G-G-F-M; $[Leu^5]$-Enkephalin has mol. wt = 556 and sequence Y-G-G-F-K		Opiate receptor sites	Possesses potent agonist activity at opiate receptor sites	Action antagonized by naloxone	4

3. REFERENCES

1. Prunty, F.T.G. and Gardiner-Hill, H. eds (1972) *Modern Trends in Endocrinology.* Butterworth, London, Vol. 4.

2. Schwatrz, T.B. ed. (1972) *Yearbook of Endocrinology.* Yearbook Medical Publishers, Chicago.

3. Margoulies, M. ed. (1969) Protein and polypeptide hormones. *Proceedings of the International Symposium, Liège, 1968.* Excerpta Medica Foundation, Amsterdam.

4. Dawson, R.M.C., Elliott, D.C., Elliott, W.H., Jones, K.M. eds (1986) *Data for Biochemical Research (3rd. edn).* Oxford University Press, Oxford.

5. Litwack, G. ed. (1972) *Biochemical Actions of Hormones (2nd edn).* Academic Press, New York, Vol. 1.

6. Fritz, I.B. ed. (1972) *Insulin Action.* Academic Press, New York.

7. Hagen, T.C., Lawrence, A.M. and Kirstein, L.. (1972) *Metab. Clin. Exp.*, **21**, 603.

8. Root, A.W. (1972) *Human Pituitary Growth Hormone.* C. C. Thomas, Springfield, IL.

9. Gual, C.H., Kastin, A.J. and Schally, A.V. (1972) *Rec. Prog. Horm. Res.*, **28**, 173, 201.

10. Papkoff, H. (1966) in *Glycoproteins* (Gottschalk, ed.). Elsevier, Amsterdam, Vol. 5, p. 532.

11. Kathan, R.H., Reichert, L.E. and Ryan, R.J. (1967) *Endocrinology*, **81**, 45.

12. Frantz, A.G., Kleinberg, D.L. and Noel, G.L. (1972) *Rec. Prog. Horm. Res.*, **28**, 229.

13. Reichlin, S., Martin, J.B., Mitnick, M., Boshans, R.L., Grimm, Y., Bollinger, J., Gordon, J. and Malacara, J. (1972). *Rec. Prog. Horm. Res.*, **28**, 229.

14. Cornell, J.S. and Pierce, J.G. (1973) *J. Biol. Chem.*, **248**, 4327.

15. Rudinger, J., Pliska, V. and Krejcí, I. (1972) *Rec. Prog. Horm. Res.*, **28**, 131.

16. McQuillan, M.T. and Trikojus, V.M. (1966) in *Glycoproteins* (Gottschalk ed.). Elsevier, Amsterdam, Vol. 5, p. 516.

17. Frieden, E. and Lipner, H. (1971) *Biochemical Endocrinology of the Vertebrates.* Prentice-Hall, Eaglewood Cliffs, NJ.

18. Bahl, O.P. (1973) in *Hormonal Proteins and Peptides* (Li, ed.). Academic Press, New York, Vol. 1, p. 171.

19. Escamilla, R.F. ed. (1971) *Laboratory Tests in Diagnosis and Investigation of Endocrine Function (2nd edn).* F. A. Davis, Philadelphia.

20. Fuchs, F. and Klopper, A. (1971) *Endocrinology of Pregnancy.* Harper & Row, New York.

21. Turner, C.D. (1971) *General Endocrinology (5th edn).* W. B. Saunders, Philadelphia.

CHAPTER 13
GROWTH REGULATORS

M. K. O'Farrell

While cell–cell interactions and cell–substratum interactions are very important in cell and tissue growth and differentiation, a fundamental role in these processes is played by soluble polypeptides collectively called growth factors or growth regulators. Many of the initial studies on the characterization and purification of individual growth factors developed from the fractionation of serum and tissue extracts that were added to defined media to obtain optimal growth of cells in tissue culture. Other growth factors have been elucidated with the development of systems in which one can study differentiation *in vitro*. Although a great deal of the detailed information concerning growth factor function comes from studies *in vitro*, there is good evidence that they do exert control over growth and differentiation *in vivo*.

Some important characteristics of polypeptide growth factors are listed in *Table 1*. Specific references are given for each growth factor, and some more comprehensive monographs are listed at the end of the chapter.

Table 1. Important characteristics of polypeptide growth factors

Growth factors	Structure and properties	Sources	Targets	Receptors	References
EGF (epidermal growth factor) and TGF-α (transforming growth factor α)	EGF and TGF-α show 40% identity; most forms about 6 kDa; synthesized as membrane-bound precursors, processed to soluble form	EGF: mainly submaxillary gland; also isolated from human urine TGF-α: embryonic tissue, many transformed cells	Cells of all three germ layers, including fibroblasts, epithelial cells, glial cells and endothelial cells	170–175 kDa protein–tyrosine kinase receptor; same receptor for EGF TGF-α and vaccinia virus growth factor, *v-erb B* product and *c-erb B/neu* product belong to same family	1–3
PDGF (platelet-derived growth factor)	Disulfide-linked dimer of A and B chains (AA, AB, BB); dimer 31 kDa; B chain is identical to *c-sis* product	Platelets, macrophages, fibroblasts, smooth muscle cells, endothelial cells	Cells of mesenchymal origin, including fibroblasts, smooth muscle cells, glial cells	α-receptor: 170 kDa β-receptor 180 kDa Both are protein–tyrosine kinases. The PDGF α-receptor binds all PDGF isoforms with high affinity; PDGF β-receptor binds PDGF-BB with high affinity and PDGF-AB with lower affinity	4, 5
IGF-I (insulin-like growth factor-I) also known as somatomedin C	7–7.6 kDa; 48% amino acid identity with pro-insulin; associates with specific binding protein in serum	Liver, lung; many cell types in culture	Cells of mesenchymal origin	Dimer of two chains, 130 kDa and 90 kDa; this receptor binds IGF-I with high affinity and IGF-II with lower affinity; protein–tyrosine kinase	6

Table 1. Continued

Growth factors	Structure and properties	Sources	Targets	Receptors	References
IGF-II (insulin-like growth factor-II) also known as somatomedin A or MSA	7–7.5 kDa; approximately 50% amino acid identity with IGF-I and insulin; associates with specific binding protein in serum	Mesenchymal cells; secretion may be developmentally regulated	Cells of mesenchymal origin; stimulates fetal development	250 kDa receptor binds only IGF-II, identical to mannose-6-phosphate receptor	6
FGF-1 (fibroblast growth factor-1) also known as acidic FGF or endothelial cell growth factor	16–17 kDa; binds strongly to heparin; 40% identity with FGF-2; member of a family of related proteins encoded by 7 distinct genes including *int-2* and *hst*	Isolated from eye and brain; also produced in the embryo	A broad range of mesoderm- and ectoderm-derived cells, including fibroblasts, smooth muscle cells, vascular endothelial cells, glial cells	150 kDa protein–tyrosine kinase receptor	7, 8
FGF-2 (fibroblast growth factor-2) also known as basic FGF	18 kDa; binds strongly to heparin; 40% identity with FGF-1; member of a family of related proteins encoded by 7 distinct genes including *int-2* and *hst*	Produced in many adult tissues and also in the embryo	A broad range of mesoderm- and ectoderm-derived cells, including fibroblasts, smooth muscle cells, vascular endothelial cells, glial cells	130 kDa protein–tyrosine kinase receptor	7, 8

Table 1. Continued

Growth factors	Structure and properties	Sources	Targets	Receptors	References
TGF-β family (transforming growth factor $\beta 1$, $\beta 2$, $\beta 3$)	25 kDa dimer; synthesized as a large precursor, released from cells as an inactive complex; other members of the TGF-β include inhibin, activin, mullerian inhibiting substance (MIS) and the decapentaplegic locus of *Drosophila*	Produced in most adult tissues and many cell types in culture; also expressed in embryos	Elicits multiple cellular responses; stimulation of proliferation is modulated via other factors; reversible growth inhibitory activity against epithelial, endothelial, fibroblast, neuronal, lymphoid and hematopoietic cell types	Three receptors: type 1, 50–60 kDa; type 2, 70–100 kDa; type 3 (betaglycan), 200–400 kDa; each type binds all forms of TGF-β	9, 10
NGF (nerve growth factor)	Dimer of 26 kDa; binds to carrier proteins in the serum	Submaxillary gland	Not a very potent mitogen but induces differentiation and enhances survival of neural cells in culture; prevents loss of neurons *in vivo*	70–80 kDa	11, 36
GRP (gastrin releasing peptide), bombesin	GRP, 3 kDa; bombesin, 14 amino acids	Neural and endocrine cells, some small-cell lung carcinomas secrete GRP; bombesin from frog skin	Fibroblasts, smooth muscle cells, neurons, small-cell lung carcinoma cells	70–90 kDa glycoprotein with 7 transmembrane sequences; G-protein-coupled receptor	12–14

Growth factors	Structure and properties	Sources	Targets	Receptors	References
Erythropoietin	34–39 kDa glycoprotein	Kidney and fetal liver	Immature erythroid cells; has some mitogenic effect but also enhances survival and prevents programmed cell death; modulates the number of circulating erythrocytes	66 kDa glycoprotein	15–17
IL-1 α and β (interleukin-1 α and β)	Both α and β species are 17 kDa; 23% identity between them	Variety of cell types including macrophages	Central mediator of inflammation; main target cells lymphocytes, epithelial cells, fibroblasts, hepatocytes, chondrocytes	Type 1 receptor, 80 kDa glycoprotein; type 2 receptor, 65 kDa; members of the immunoglobulin family	18, 19
IL-2 (interleukin-2)	15 kDa	Activated T cells	Induces the growth and clonal expansion of primary T cells; long-term maintenance and growth of T-cell lines; also influences B-cell proliferative responses	High-affinity receptor is composed of 70 kDa polypeptide and a 55 kDa polypeptide, although both possess IL-2 binding sites	20
IL-3 (interleukin-3) (also multi-CSF)	14–28 kDa glycoprotein	Antigen- or mitogen-activated T cells and some cell lines	Stimulation of most hemopoietic lineages; self-renewal, survival and differentiation of multipotential stem cells	140 kDa protein–tyrosine kinase	21–23

Table 1. Continued

Growth factors	Structure and properties	Sources	Targets	Receptors	References
IL-4 (interleukin-4)	15–20 kDa glycoprotein	Activated T cells	Stimulates proliferation and Ig production in B cells; also affects T cells, macrophages and mast cells	145 kDa glycoprotein; a smaller form acts as a soluble receptor and inhibitor of IL-4 activity	24, 25
IL-5 (interleukin-5)	15–22 kDa glycoprotein	Activated T cells	Supports the proliferation and differentiation of eosinophils; also interacts with B and T cells		26
IL-6 (interleukin-6)	19–30 kDa glycoprotein	T cells, fibroblasts, endothelial cells, macrophages, B- and T-cell lines	B and T cells, nerve cells, hepatocytes, fibroblasts and epithelial cells; used to support the growth of hybridomas	80 kDa ligand binding glycoprotein and a 130 kDa signal transducing glycoprotein	27, 28
IL-7 (interleukin-7)	25 kDa glycoprotein	Bone marrow stromal cells	Stimulates pre-B-cell proliferation; also has stimulatory effects on mature T cells and thymocytes	65–75 kDa glycoprotein	29, 30

Table 1. Continued

Growth factors	Structure and properties	Sources	Targets	Receptors	References
CSF-1 (colony-stimulating factor) known as M-CSF (macrophage colony-stimulating factor)	Homodimer of 2 chains 14–21 kDa, linked by disulfide bonds; glycosylated	Multiple cell types including fibroblasts, endothelial cells, monocytes and macrophages; a variety of tumor-derived cells also produce CSF-1	Stimulates proliferation and differentiation of bone-marrow progenitor cells to form macrophages; required for monocyte and macrophage survival; also has a role in placental development	150 kDa glycoprotein with protein–tyrosine kinase activity; product of the *c-fms* proto-oncogene	31–33
G-CSF (granulocyte colony-stimulating factor)	24–25 kDa glycoprotein	Multiple cell types including fibroblasts, endothelial cells and macrophages; also produced by several tumor-derived cell lines	Stimulates production of granulocytes; also induces terminal differentiation of granulocytes	150 kDa	31, 34
GM-CSF (granulocyte-macrophage colony-stimulating factor)	15–30 kDa glycoprotein	Multiple cell types including activated T lymphocytes, fibroblasts, endothelial cells and macrophages	Stimulates production of neutrophilic and eosinophilic granulocytes and macrophages; initiates proliferation of erythroid and megakaryocyte precursors	80 kDa α-subunit with low ligand-binding affinity associates with a 120 kDa β polypeptide to form a high-affinity complex; both polypeptides are glycosylated	31, 35

Table 1. Continued

Growth factors	Structure and properties	Sources	Targets	Receptors	References
TNF-α (cachectin)	Trimer with 17 kDa subunits; 30% identity with TNF-β	Predominantly stimulated macrophages	Stimulates proliferation of cultured fibroblasts and some tumor cell lines; cytoxic for some transformed cells	55 kDa receptor binds TNF-α and TNF-β with equal affinity; a second receptor of 68–75 kDa has been isolated	37, 38
TNF-β (lymphotoxin)	25 kDa glycosylated polypeptide; 30% identity with TNF-α	T lymphocytes		55 kDa receptor binds TNF-α and TNF-β with equal affinity	39
IFN-α (interferon-α)	20–25 kDa; 30% amino acid identity with IFN-β	Most cell types; induced by viruses, dsRNA and growth factors and other cytokines	Inhibits proliferation in a number of cell types, including transformed and tumor cell lines	95–100 kDa	40–42
IFN-β (interferon-β)	25–35 kDa glycoprotein; 30% amino acid identity with IFN-α	Most cell types; induced by viruses, dsRNA and growth factors and other cytokines	Inhibits proliferation in a number of cell types, including transformed and tumor cell lines	Shares the receptor for IFN-α	40, 41
IFN-γ (interferon-γ)	20–25 kDa glycoprotein	T cells	Weak antiproliferative effect	117 kDa protein	40, 41, 43

1. SPECIFIC REFERENCES

1. Carpenter, G and Cohen, S. (1990). *J. Biol. Chem.*, **265**, 7709.

2. Todaro, G.T., Rose, T.R., Spooner, C.E., Shoyab, M. and Plowman, G.D. (1990) *Seminars in Cancer Biology*, **1**, 257.

3. Kudlow, J.E. and Bjorge, J.D. (1990) *Seminars in Cancer Biology*, **1**, 293.

4. Ross, R., Bowen-Pope, D.F. and Raines, E.W. (1990) *Phil. Trans. Roy. Soc. (Lond.), Series B*, **327**, 155.

5. Heldin, C.-H. and Westermark, B. (1990) *Cell Regul.*, **1**, 555.

6. Le Roith, D. and Raizada, M.K. (eds) (1989) *Molecular and Cellular Biology of Insulin-like Growth Factors and their Receptors.* Plenum, New York.

7. Burgess, W.H. and Maciag, T. (1989) *Ann. Rev. Biochem.*, **58**, 575.

8. Goldfarb, M. (1990) *Cell Growth and Differentiation*, **1**, 439.

9. Massague, J. (1990) *Ann. Rev. Cell Biol.*, **6**, 597.

10. Barnard, J.A., Lyons, R.M. and Moses, H.L. (1990) *Biochim. Biophys. Acta*, **1032**, 79.

11. Hempstead, B.L. and Chao, M.S. (1989) *Recent Progress in Hormone Res.*, **45**, 441.

12. Rozengurt, E. and Sinnett-Smith, J. (1983) *Proc. Natl. Acad. Sci. USA*, **80**, 2936.

13. Cuttitta, F., Carney, D.N., Mulshine, J., Moody, T.W., Fedorko, J., Fischler, A. and Minna, J.D. (1985) *Nature*, **316**, 823.

14. Battey, J.F., Way, J.M., Corjay, M.H., Shapira, H., Kusano, K., Harkins, R., Wu, J.M., Slattery, T., Mann, E. and Feldman, R.I. (1991) *Proc. Natl. Acad. Sci. USA*, **88**, 395.

15. McDonald, J.D., Lin, E.-K. and Goldwasser, E. (1986) *Mol. Cell. Biol.*, **6**, 842.

16. Jones, S.S., D'Andrea, A.D., Haines, L.L. and Wong, G.G. (1990) *Blood*, **76**, 31.

17. Koury, M.J. and Bondurant, M.C. (1990) *Science*, **248**, 378.

18. Dower, S.K. and Sims, J.E. (1990) in *Cellular Molecular Mechanisms of Inflammation* (C.G. Cochrane and M. Gimbrone eds). Academic Press, London, Vol. 1, p. 137.

19. Dinarello, C.A. (1991) in *The Cytokine Handbook* (A.W. Thomson ed.). Academic Press, London, p. 47.

20. Kuziel, W.A. and Greene, W.C. (1991) in *The Cytokine Handbook* (A.W. Thompson ed.). Academic Press, London, p. 83.

21. Schrader, J.W. (1991) in *The Cytokine Handbook* (A.W. Thomson ed.). Academic Press, London, p. 103.

22. Itoh, N., Yonehara, S., Schreurs, J., Gorman, D.M., Maruyama, K., Ishii, A. *et al.* (1990) *Science*, **247**, 324.

23. Whetton, A.D. and Dexter, T.M. (1989) *Biochim. Biophys. Acta*, **989**, 111.

24. Harada, N., Castle, B.E., Gorman, D.M., Itoh, N., Schreurs, J., Barrett, R.L., Howard, M. and Miyajima, A. (1990) *Proc. Natl. Acad. Sci. USA*, **87**, 857.

25. Banchereau, J. (1991) in *The Cytokine Handbook* (A.W. Thomson ed.). Academic Press, London, p. 119.

26. Sanderson, C.J. (1991) in *The Cytokine Handbook* (A.W. Thomson ed.). Academic Press, London, p. 146.

27. Sehgal, PB., Greininger, G. and Tosato, G. (eds). (1989) Regulation of the acute phase and immune responses: interleukin 6., *Ann. NY Acad. Sci*, **57**.

28. Hirano, T. (1991) in *The Cytokine Handbook* (A.W. Thomson ed.). Academic Press, London, p. 169.

29. Cosman, D., Lyman, S.D., Idzerda, R.L., Beckmann, M.P., Park, L.S., Goodwin, R.G. and March, C.J. (1990) *TIBS*, **15**, 256.

30. Goodwin, R.G. and Namen, A.E. (1991) in *The Cytokine Handbook* (A.W. Thomson ed.). Academic Press, London, p. 191.

31. Garland, J.M. (1991) in *The Cytokine Handbook* (A.W. Thomson ed.). Academic Press, London, p. 269.

32. Stanley, E.R. (1990) in *Genetics of Pattern Formation and Growth Control* (A.P. Mahowald ed.). Society for Developmental Biology, Berkeley, p. 165.

33. Sherr, C.J. (1990) *Blood*, **75**, 1.

34. Fukunaga, R., Ishizaka-Ikeda, E., Seto, Y. and Nagata, S. (1990) *Cell*, **61**, 341.

35. Gough, N.M. and Nicola, N.A. (1990) in *Colony Stimulating Factors: Molecular and Cellular Biology* (T.M. Dexter, J.M. Garland and N.G. Testa eds). Marcel Dekker, New York, p. 111.

36. Chao, M.V. (1989) in *Growth Factors, Differentiation Factors and Cytokines* (A. Habenicht ed.). Springer-Verlag, Berlin, p. 65.

37. Le, J. and Vileck, J. (1987) *Lab. Invest.*, **56**, 234.

38. Manogue, K.R., van Deventer, S.J.H. and Cerami, A. (1991) in *The Cytokine Handbook*

(A.W. Thomson ed.). Academic Press, London, p. 241.

39. Ruddle, N.H. and Turetskaya, R.L. (1991) in *The Cytokine Handbook* (A.W. Thomson ed.). Academic Press, London, p. 257.

40. Pestka, S., Langer, J.A., Zoon, K.C. and Samuel, C.E. (1987) *Ann. Rev. Biochem.*, **56**, 727.

41. de Maeyer, E. and de Maeyer-Guignard, J. (1991) in *The Cytokine Handbook* (A.W. Thompson ed.). Academic Press, London, p. 215.

42. Uzes, G., Lutfalla, G. and Gresser, I. (1990) *Cell*, **60**, 225.

43. Aguet, M., Dembric, Z. and Merlin, G. (1988) *Cell*, **55**, 273.

2. GENERAL REFERENCES

Arai, K., Lee, F., Miyajima, A., Miyatake, S., Arai, N. and Yakota (1990) *Ann. Rev. Biochem.*, **59**, 783.

Deuel, T.F. (1987) *Ann. Rev. Cell Biol.*, **3**, 443.

Dexter, T.M., Garland, J.M. and Testa, N.G. (eds) (1990) *Colony Stimulating Factors*. Marcel Dekker, New York.

Habenicht, A. (ed.) (1989) *Growth Factors, Differentiation Factors and Cytokines*. Springer-Verlag, Berlin.

Jenkins, N. (1991) in *Mammalian Cell Biotechnology* (M. Butler ed.). IRL Press, Oxford.

LeRoith, D. and Raizada, M.K. (eds) (1989) *Molecular and Cellular Biology of Insulin-like Growth Factors and their Receptors*. Plenum, New York.

Mercola, M. and Stiles, C.D. (1988) *Development*, **102**, 451.

Sporn, M.B. and Roberts, A.B. (eds) (1990) *Handbook of Experimental Pharmacology: Peptide Growth Factors and their Receptors*. Springer-Verlag, Berlin, Part 1, Vol. 95.

Thomson, A.W. (ed.) (1991) *The Cytokine Handbook*, Academic Press, London.

Waterfield, M.D. (ed.), (1989) Growth factors. *British Medical Bulletin*, **45**.

Whitman, M. and Melton, D.A. (1989) *Ann. Rev. Cell Biol.*, **5**, 93.

CHAPTER 14

THE CELL CYCLE: DETERMINATION OF CELL CYCLE PARAMETERS

M. K. O'Farrell and G. B. Dealtry

1. DETERMINATION OF GROWTH FRACTION

The proportion of cells in a cell population that are actively proliferating is the growth fraction. This parameter can be measured in cell populations *in vivo* and *in vitro*; usually using continuous labeling with [^{3}H]thymidine to determine the maximum number of cells that are able to synthesize DNA. Cells are incubated with [^{3}H]thymidine for long periods of time (12, 24, 36 or 48 h) depending on the estimated generation time of the cell population. Following fixation and autoradiography, the percentage of labeled nuclei is determined and the maximum value obtained represents the growth fraction (1). Underestimation of the proportion of non-cycling cells can arise if the labeling period is long, and thus picks up cells entering the S phase for a second time. Bromodeoxyuridine incorporation, followed by detection with specific antibody, can be used as an alternative to radioactive thymidine. In exponentially growing permanent eucaryotic cell lines in culture the growth fraction should be close to 100%. In non-established cell strains the growth fraction will change with passage number.

2. DETERMINATION OF TOTAL CELL CYCLE TIME AND THE LENGTH OF INDIVIDUAL PHASES, G1, S, G2 AND M

The cell cycle, or generation time, is the time taken to proceed from one cell division to the next. The average cell cycle time in a cell population is frequently measured as the time taken to achieve double the population number; this is valid only if the growth fraction is 100%. The time between divisions can be determined using time-lapse cinematography or video microscopy (2, 3). The cell cycle of eucaryotic cells has been divided into discrete phases; G1 (G0), S, G2 and M. The total cell cycle time (Tc), as well as the length of individual phases, can be measured using the frequency (or fraction) of labeled mitosis, or FLM. A cohort of S-phase cells are labeled with a brief pulse of [^{3}H]thymidine and are analyzed as they pass through mitosis, a microscopically recognizable point in the cell cycle. The time taken to transit G2, M, S can be obtained directly from the data collected, as can the total cell cycle time (Tc). The length of the G1 phase is determined by difference (4, 5). The duration of mitosis and cytokinesis is not determined accurately by this method; time-lapse cinematography gives a more accurate estimation (6).

A simpler method involving labeling of the cell population with [^{3}H]thymidine for progressively longer periods of time allows the determination of the rate of entry into the S-phase. This is assumed to be the same as the rate of entry into, and therefore the rate of exit from, cell division; so the rate of cell division can be determined. This method also gives a measure of the fraction of cells in the S phase (1).

Flow cytometry can also be used to determine the proportion of cells in the population in different phases of the cell cycle. Cells are treated with fluorescent dyes, such as propidium iodide, that bind specifically to DNA; the quantity of fluorescence is a measure of the amount of DNA per cell. Because the DNA doubles during the S phase, the DNA content of G2-phase cells is twice that of G1 cells. Using flow cytometry the fluorescence, and therefore the DNA content, of a large number of cells can be measured and plotted as a histogram. This characteristically shows two peaks, representing the G1 and G2 complement of DNA, separated by a trough representing cells with intermediate quantities of DNA progressing through the S phase. The fraction of cells in G1, S and G2 + M can be obtained by computer-fit analysis; however, it is important to recognize that this in itself does not yield values of duration. These can only be obtained if one has determined Tc by an independent method (7, 8).

3. SYNCHRONIZATION OF CELL CYCLE

In order to study the biochemistry and molecular biology of the cell cycle, it is useful to have populations of cells at the same stage of the cell cycle. While it is possible to study the cell cycle in naturally synchronous systems (for example, sea-urchin eggs and the acellular slime mold *Physarum polycephalum*), proliferation in most cell populations is asynchronous. There are two general types of experimentally derived synchrony, induction synchrony and selection synchrony. When choosing a synchronization protocol it is important to consider several factors:

(i) the degree of perturbation to the metabolic and physiological state of the cell;
(ii) the ease of reversibility (when using an inhibitor);
(iii) the degree of synchrony obtained; and
(iv) the suitability of a particular method for studying a specific part of the cell cycle.

Induction of synchrony involves blocking the cells at a defined, periodic event in the cell cycle, usually with metabolic inhibitors. One possible site for intervention is DNA replication. A frequently used method is that of the double thymidine block. High concentrations of thymidine inhibit DNA replication by inhibiting nucleoside diphosphate reductase, and when two periods of thymidine inhibition separated by a carefully timed release period are applied a reasonable degree of synchrony is obtained (9). Other inhibitors of DNA synthesis used to induce synchrony include fluorodeoxyuridine, hydroxyurea, cytosine arabinoside and aphidicolin (9, 10, 13). Usually they have to be applied as a double metabolic block or in combination with another form of arrest, for example growth factor deprivation or mitotic selection (11, 12).

Enrichment of mitotic cells can be achieved using inhibitors of mitosis, for example colchicine, vinblastine or taxol (25).

There are also several ways in which cell populations can be synchronized in the G1 phase. Mammalian cells can be arrested at a point in G1 by depletion of the amino acids isoleucine or leucine in the culture medium; readdition of the amino acid allows continued progression through the cycle (14, 15). Normal animal cells in culture become arrested in G1 and enter into a quiescent G0 state when cultures are allowed to reach confluence. This probably represents a combination of density inhibition of proliferation and growth factor deprivation. Cells can be restimulated into the cycle by addition of fresh serum or purified growth factors. This experimental situation is most useful for the study of the exit from G0 and of the events leading up to the initiation of DNA synthesis (16, 17).

With increasing interest in the molecular events controlling the entry into mitosis, efforts have been made to identify inhibitors for use in establishing G2 synchrony. Inhibitors of protein kinases and of topoisomerase II have been used to obtain synchrony (15, 18).

Several other methods of induction synchrony have been developed for particular cell types. For example, heat shock has been used to synchronize cell divisions in the cilliate *Tetrahymena* (19). Many unicellular algae can be synchronized by periodic darkening at intervals related to the diurnal cycle. Darkness stops photosynthesis and so halts cells in the growth phase of the cycle. Repeated light–dark cycles yield highly synchronized populations in *Chlamydomonas* and *Chlorella* (20).

Selection synchrony methods usually involve substantially less perturbation than those using induction processes. They consist primarily of separation of cell populations on the basis of size and, to a lesser extent, density. Centrifugal elutriation has been used very successfully to obtain analytical quantities of cells at different stages of the cell cycle with minimal metabolic perturbation (21, 23). This technique has been used with yeasts (22) as well as with higher eucaryotic cells (23). Cell cycle fractions of high degrees of purity can be obtained; for example up to 95% in G1- and S-phase populations. The other advantage is that it is possible to process large numbers of cells (up to 3×10^9 cells).

As monolayer cells enter mitosis they become more spherical in shape and attach less tightly to the substratum. The mitotic cells can be detached selectively from the monolayer and put back into culture to study G1 and, with a lower degree of synchrony, S-phase events. A drawback of this method is that the yield of mitotic cells is rather low and thus large areas of monolayer culture are necessary to obtain enough cells for biochemical analysis (9, 24, 25, 26).

Although flow cytometry is used most frequently as an analytical technique, it can, with the appropriate machines, be extended to include cell sorting. Cell sorting provides cell suspensions of a high degree of purity for the particular parameter chosen. Using light scatter one can select viable cell populations of particular mass and, provided the sterile conditions are maintained, culture these cells through the cycle. A limitation to this technique is again likely to be one of cell number (27, 28).

Table 1. Cell cycle times of some eucaryotic and procaryotic cells[a]

Species	**Cycle time (Tc) (h)**
Aspergillus nidulans	2
Physarum polycephalum	10
Tetrahymena pyriformis	3
Euplotes eurystomus	14
Amoeba proteus	48
Schizosaccharomyces pombe	4
Saccharomyces cerevisiae	3
Chlorella pyrenoidosa	24
Escherichia coli	0.5
Caulobacter crescentus	
stalked cell	2
swarm cell	3

[a] Data compiled from refs 29–33.

4. CELL CYCLE DATA FOR DIFFERENT CELLS

4.1. Total cycle time (Tc)

The length of an individual cell cycle varies even within cells of the same type, depending on physiological conditions such as cell age, the presence of different nutrients and growth factors, etc. This is particularly variable in mammalian cells, where even homogeneous populations in defined tissue culture conditions show variations in individual cell cycle times. Further variables are introduced by the many different techniques employed to measure cell cycle time and to synchronize cell populations experimentally (see Sections 2 and 3). However, some representative cell cycle times are listed in *Tables 1*, *2* and *3* to provide an indication of the duration of typical cell cycles.

The cell cycle of eucaryote cells has been divided into discrete phases; G1 (G0), S, G2 and M. G1 forms the period between the end of the preceeding mitosis and the DNA synthesis (S) phase. It is the most variable phase of the cell cycle, ranging from almost no time to prolonged periods of apparent arrest when the cell enters a quiescent or resting G0 phase. S phase is reasonably constant, lasting approximately 7–10 h in higher eucaryotes,

Table 2. Typical mammalian cell cycle times *in vivo*[a]

Species	Tissue type	Cycle time (Tc) (h)
Man (normal)	Bone marrow cells	18
	Bronchus, epithelial cells	220
	Colon, crypt cells	448
Man (tumor)	Acute myeloblastic leukemia	49
	Bronchus, carcinoma	196–260
	Epidermoid carcinoma	24
	Melanoma	46
Mouse (normal)	Antibody-forming plasma cells	9
	Colon, epithelial cells	19
	Duodenal crypt cells	10.3
	Embryo ependymal cells	11
	Growing hair follicles	12
	Mammary gland, alveoli	71
	Mammary gland, alveoli after hormonal stimuli	13
Mouse (tumor)	B16 melanoma	16.5
	Lewis lung carcinoma	17.6
Rat (normal)	Internal enamel epithelium	27.3
	Liver cells (8 weeks old)	47.5
	Spleen germinal centers	13.4
Rat (tumor)	Hepatoma cells	24

[a] Data compiled from refs 34 and 35.

Table 3. Typical mammalian cell cycle times *in vitro*

Species	Cell line/type	Cycle time (Tc) (h)
Man	WI-38 diploid fetal lung fibroblasts	24.5
	amnion cells	19.5
	HeLa S3 cells	21
Mouse	3T3 cells	19
	L cells	23
Hamster	Chinese hamster ovary cells	14
	Chinese hamster lung cells	10

although there are exceptions. In yeasts S phase is reduced to approximately 40 min. The second gap phase, G2, is the period between DNA synthesis and the onset of mitosis, and, like G1, is quite variable in duration. Mitosis usually lasts less than an hour, although again there are exceptions.

The breakdown of the cell cycles of a number of actively replicating mammalian cells in culture are given in *Table 4*, as an indication of the durations of the phases.

Table 4. Length of phases of cell cycle in selected mammalian cells in culture[a]

Cell type	G1 (h)	S (h)	G2 + M (h)	Tc (h)
Human amnion cells	9.5	7.0	3.0	19.5
Human diploid fibroblasts	6.5	7.5	4.0	18
HeLa S3	8.0	9.5	3.5	21
KB human oral epidermal cells	6.5	7.5	17.5	31.5
Mouse L fibroblasts	8.0	6.0	4.0	18
Mouse 3T3	8.0	7.0	4.0	19
Chinese hamster lung cells	1.5	6.0	2.5	10

[a]Data derived from ref. 34.

5. REFERENCES

1. Baserga, R. (1989) in *Cell Growth and Division: A Practical Approach* (R. Baserga ed.). IRL Press/Oxford University Press, Oxford, p. 1.

2. Shields, R., Brooks, R.F., Riddle, P.N., Capellaro, D.F. and Delia, D. (1978) *Cell*, **15**, 469.

3. Zetterberg A. and Larsson, O. (1985) *Proc. Natl Acad. Sci. USA*, **82**, 5365.

4. Steel, G.G. and Hanes, S. (1971) *Cell Tiss. Kinet.*, **4**, 93.

5. Karatza, C., Stein, W.D., and Shall, S. (1984) *J. Cell Sci.*, **65**, 163.

6. Sisken, J.E. and Wilkes, E. (1967) *J. Cell Biol.*, **34**, 97.

7. Gray, J.W., Dolbeare, F. and Pallavicini, M. G. (1990) in *Flow Cytometry and Sorting (2nd edn.)* (M.R. Melamed, T. Lindmo and M.L. Mendelsohn eds). Wiley-Liss, New York, p. 445.

8. Watson, J.V. (1991) *Introduction to Flow Cytometry.* Cambridge University Press, Cambridge.

9. Stein, G.S. and Stein, J.L. (1989) in *Cell Growth and Division: A Practical Approach* (R. Baserga ed.). IRL Press/Oxford University Press, Oxford, p. 133.

10. Grdina, D.J., Meistrich, M.L., Meyn, R.E., Johnson, T.S. and White, R.A. (1987) in *Techniques in Cell Cycle Analysis* (J.W. Gray and Z. Darynkiewicz eds). Humana Press, p. 367.

11. Gurley, R.L., Waters, R.A. and Tobey, R.A. (1973) *Biochem. Biophys. Res. Commun.*, **50**, 744.

12. Cao, G., Liu, L.-M. and Cleary, S. (1991) *Exp. Cell Res.*, **193**, 405.

13. Pedrali-Roy, G., Spadari, S., Miller-Faures, A., Miller, A.O.A., Kruppa, J. and Koch, G. (1980) *Nucleic Acids Res.*, **8**, 377

14. Ley, K.D. and Tobey, R.A. (1970) *J. Cell Biol.*, **47**, 453.

15. Tobey, R.A., Oishi, N. and Crissman, H.A. (1990) *Proc. Natl. Acad. Sci. USA*, **87**, 5104.

16. Jimenez de Asua, L., O'Farrell, M.K., Clingan, D. and Rudland, P.S. (1977) *Proc. Natl. Acad. Sci. USA*, **74**, 3845.

17. Scher, C.D., Shephard, R.C., Antoniades, H.N. and Stiles, C.D. (1979) *Biochim. Biophys. Acta.*, **560**, 217.

18. Abe, K., Yoshida, M., Usui, T., Horinouchi, S. and Beppu, T. (1991) *Exp. Cell. Res.*, **192**, 122.

19. Everhart, L.P. (1972) in *Methods in Cell Biol.* (D.M. Prescott ed.). Academic Press, London, Vol. 5, p. 219.

20. John, P.C.L., Lambe, C.A., McGookin, R. and Orr, B. (1981) in *The Cell Cycle.* Society for Experimental Biology, Seminar Series 10. Cambridge University Press, Cambridge, p. 185.

21. D. Conkie (1986) in *Animal Cell Biology: A Practical Approach* (R.I. Freshney ed.). IRL Press, Oxford, p. 113.

22. Aves, S.J., Durkacz, B.W., Carr, A. and Nurse, P. (1985) *EMBO J.*, **4**, 457.

23. Kauffman, M.G., Noga, S.J., Kelly, T.J. and Donnenberg, A.D. (1990) *Anal. Biochem.*, **191**, 41.

24. Terasima, T. and Tolmach, L. (1963) *Exp. Cell Res.*, **30**, 344.

25. Gaffney, E.V. and McElwain, E.G. (1973) *In Vitro*, **9**, 56.

26. Lesser, B. and Brent, T.P. (1972) *Exp. Cell Res.*, **62**, 470.

27. Borysiewicz, L.K. (1990) in *Flow Cytometry: A Practical Approach* (M.G. Ormerod ed.). IRL Press, Oxford.

28. Hoy, T.G. (1990) in *Flow Cytometry: A Practical Approach* (M.G. Ormerod ed.). IRL Press, Oxford.

29. Mitchison, J.M. (1971) *The Biology of the Cell Cycle.* Cambridge University Press, Cambridge.

30. Prescott, D.M. (1976) *Reproduction of Eukaryotic Cells.* Academic Press, London.

31. John, P.C.L. (ed.) (1981) *The Cell Cycle.* Society for Experimental Biology, Seminar Series 10. Cambridge University Press, Cambridge.

32. Lloyd, D., Poole, R.K. and Edwards, S.W. (eds) (1982) *The Cell Division Cycle — Temporal Organisation and Control of Cellular Growth and Reproduction.* Academic Press, London.

33. Cooper, S. (1991) *Bacterial Growth and Division.* Academic Press, London.

34. Baserga, R. (1985) *The Biology of Cell Reproduction.* Harvard University Press, Cambridge, Mass.

35. Baserga, R. (1976) *Multiplication and Division in Mammalian Cells.* Marcel Dekker, New York.

CHAPTER 15

ONCOGENES

M. K. O'Farrell

Oncogenes can be broadly defined as a class of genes whose inappropriate expression or expression in an altered form can lead to neoplastic transformation. Generally, the proto-oncogenes from which they are derived have been shown to be involved in the control of cellular proliferation or cellular differentiation.

Oncogenes have been identified using a number of different approaches. The analysis of the transforming genes in the acutely transforming retroviruses has led to the characterization of many transduced cellular sequences that are oncogenic when expressed by the virus. Tumors can also be caused by insertion of retroviruses into the host cell genome; this leads to the activation of adjacent cellular genes. A number of putative proto-oncogenes have been identified in this way. DNA transfer technologies have been important in identifying activated oncogenes from human tumors. While this technique seems to be fairly limiting in the class of oncogene that it detects (at least with the commonly used NIH 3T3 fibroblasts as recipient), a number of novel oncogenes generated by genetic rearrangement have been detected. The characterization of chromosomal translocations specific for particular malignancies as well as specific DNA amplifications has also been very important in the identification of oncogenes.

The tables below bring together genetic, cell biological and biochemical information about a wide range of well-studied oncogenes and proto-oncogenes. New oncogenes/proto-oncogenes continue to be discovered, so a collection of data like this will inevitably be incomplete.

Table 1. Oncogenes and proto-oncogenes

Proto-oncogene	Viral or tumor origin	Protein product	Cellular location of product	Function of protein	Human chromosome location	Tissue expression	Other information
abl	Abelson murine leukemia virus	*c-abl* 145–150 kDa protein *gag-abl* 160 kDa phosphoprotein *bcr-abl* 210 kDa phosphoprotein	Predominantly soluble and membrane fractions of the cytoplasm; some inner face of plasma membrane	Non-receptor protein-tyrosine kinase	9q34.1	Virtually all tissues	Activation of the *c-abl* gene can occur through incorporation into a retrovirus and by chromosomal translocation; virus induces primarily pre-B-cell tumors; the translocated *abl* proto-oncogene has been implicated in the cause and maintenance of certain forms of human leukemia that carry the Philadelphia chromosome (1, 2)

Table 1. Continued

Proto-oncogene	Viral or tumor origin	Protein product	Cellular location of product	Function of protein	Human chromosome location	Tissue expression	Other information
bcl-2	Chromosome translocation in follicular lymphoma	25 kDa protein	Cell membrane	—	18q21.3	Lymphoid cells	Isolated based on its rearrangement to the immunoglobulin IGH gene on chromosomal translocation; translocation leads to higher than normal levels of mRNA and protein due to transcriptional deregulation; *bcl* codes for an integral membrane protein (3–5)
cbl	Murine Cas-Br-M virus	*c-cbl* 135 kDa protein *gag-cbl* 100 kDa protein	Nucleus	Possible transcription factor	11q23	Most mouse tissues; many human and mouse tumors of hemopoietic tissues	Virus induces predominantly pre-B-cell lymphomas and *v-cbl* causes transformation of fibroblasts *in vitro*; sequence analysis indicates a putative DNA-binding domain (6)
cot	Human thyroid carcinoma cell line, TC04	52 kDa protein	Primarily cytoplasmic	Protein-serine/threonine kinase	—	—	Overexpression of *c-cot* transforms mouse NIH3T3 cells and hamster cells (7)

Table 1. Continued

Proto-oncogene	Viral or tumor origin	Protein product	Cellular location of product	Function of protein	Human chromosome location	Tissue expression	Other information
crk	Avian sarcoma virus C10	*c-crk* 34 kDa phosphoprotein *crk-gag* 47 kDa protein	Cytoplasmic	Possible regulator of protein-tyrosine kinase	—	Most embryonic and adult tissues	Activated by viral transduction; *v-crk* contains sequences related to regulatory sequences called SH-2 and SH-3 found in the non-receptor protein-tyrosine kinases but this oncogene lacks a protein kinase catalytic domain; *v-crk* binds to protein-tyrosine kinases and other proteins through phospho-tyrosine present at phosphorylation sites (8–10)
dbl	Diffuse B-cell lymphoma	*c-dbl* 115 kDa protein Oncogene 66 kDa protein	Cytoplasmic; soluble and membrane bound with some evidence of association with the cytoskeleton	—	Xq27–28	Fetal brain and adrenal; adult testes and ovaries	Activation probably occurred by gene rearrangement during DNA transfection (11, 12)

Table 1. Continued

Proto-oncogene	Viral or tumor origin	Protein product	Cellular location of product	Function of protein	Human chromosome location	Tissue expression	Other information
erbA	Avian erythroblastosis virus-ES4 and -R strains	*gag-erbA* 75 kDa protein *c-erbA/TR* 46 kDa protein	Nucleus	High-affinity nuclear receptor for thyroid hormone	17p11–q21	—	Activation by viral transduction; both *c-erbA* and *v-erbA* bind DNA with high affinity but *v-erbA* is defective in binding thyroid hormone; a second *c-erbA*-related gene, *c-erbA*β, mapping to chromosome 3, has been found in the human genome; no rearrangement or enhanced expression of either *c-erbA/TR* loci have yet been correlated with human disease (13, 14)
erbB1	Avian erythroblastosis virus-ES and -R strains	*v-erbB* 75 kDa glycoprotein *c-erbB EGFR* 170 kDa glycoprotein	Plasma membrane	Truncated EGF receptor protein tyrosine kinase	7p11–q22	Most cell types; hematopoietic cells do not normally express the gene product	Activated by viral transduction and promoter insertion; the *v-erbB* protein is homologous to the transmembrane and cytoplasmic domains of the EGF receptor. The *c-erbB EGFR* locus is amplified in several human tumors (15–17)

Table 1. Continued

Proto-oncogene	Viral or tumor origin	Protein product	Cellular location of product	Function of protein	Human chromosome location	Tissue expression	Other information
ets	Avian myeloblastosis virus	*gg-myb-ets* 135 kDa protein *c-ets* (chicken) 54 kDa phosphoprotein *c-ets1* (human) 40–51 kDa phosphoproteins *c-ets2* (human) 56 kDa phosphoprotein	*c-ets* (chicken) is largely cytoplasmic but also probably resides in the nucleus *c-ets1* and *c-ets2* (human) are both nuclear proteins	Sequence-specific DNA-binding proteins; transcriptional regulators?	*c-ets1* 11q23–q24 *c-ets2* 21q22.3	*c-ets* (chick) and *c-ets1* (human) are enriched in lymphoid tissues; *c-ets2* is expressed in a wide variety of tissues; expression linked to cell proliferation	Overexpression of mouse *c-ets2* transforms NIH 3T3 cells; the genes are members of a gene family which include *elk1*, *elk2* and *erg* (18–21)
evi-1	Site of viral integration (ecotropic virus integration site 1) in mouse myeloid leukemia cell lines	145 kDa zinc-finger protein	Nucleus	Sequence-specific DNA-binding; transcription factor?	—	Renal tubules in the corticomedullary junction; developing oocytes in the ovary; not expressed in normal hemopoietic cells	Retroviral insertions in the *Cb-1/fim-3* locus also activate the expression of the *evi-1* gene (22–25)
evi-2	Site of viral integration (ecotropic virus integration site 2) in	cDNA clones indicate 24 kDa transmembrane protein	Cell membrane?	—	17q11.2	Mouse brain, ovaries and macrophages	Two other genes have been found in the region of *evi-2*, *evi-2b* and the gene responsible for the

Table 1. Continued

Proto-oncogene	Viral or tumor origin	Protein product	Cellular location of product	Function of protein	Human chromosome location	Tissue expression	Other information
	mouse myeloid tumors						human disease neurofibromatosis type 1 (Nf1); as the Nf1 gene shows homology to two yeast genes that are involved in the regulation of RAS, it is not yet known which of these genes is important in the insertional event (26–28)
fgr	Gardner–Rasheed feline sarcoma virus	*gag-actin-fgr* 70 kDa protein *c-fgr* 55 kDa protein	Plasma-membrane associated	Non-receptor protein-tyrosine kinase	1p36.2–36.1	Macrophages, monocytes, granulocytes	Activated by viral transduction also may be induced by EB virus infection/transformation; *c-fgr* belongs to the *src* family of protein-tyrosine kinases (29–31)
fms	Susan McDonough feline sarcoma virus (SM-FeSv)	*gag-fms* 180 kDa glycoprotein *c-fms* 120 kDa glycoprotein	Plasma membrane	Colony stimulating factor-1 (CSF-1) receptor protein-	5q34	Bone marrow and peripheral blood mono-nuclear cells; also	Insertional activation of the *c-fms* gene has been implicated as an early step leading to the development of Friend murine

Table 1. Continued

Proto-oncogene	Viral or tumor origin	Protein product	Cellular location of product	Function of protein	Human chromosome location	Tissue expression	Other information
				tyrosine kinase		expressed in choriocarcinoma cell lines	leukemia virus-induced leukemia; *v-fms* retains the complete extracellular domain of the *c-fms* gene product and specifically binds CSF-1 (32–34)
fos	Finkel–Biskis–Jinkins murine osteosarcoma virus (FBJ-MSV) and Finkel–Biskis–Reilly murine osteosarcoma virus (FBR-MSV)	*v-fos* (FBJ-MSV) 55 kDa phosphoprotein *gag-fos-fox* (FBR-MSV) 75 kDa protein *c-fos* 62 kDa phosphoprotein	Nucleus	Binds with members of the *jun* gene family to form heterodimers that interact with DNA at AP1 sites; *trans*-acting gene regulator	14q21–31	Expressed at low levels in most growing cells; amnion, yolk sac, fetal liver, placenta and postnatal bone marrow cells express fairly high levels	Activation occurs by the addition of a transcriptional enhancer sequence and the disruption of an interaction between the gene-coding sequence and a 3′ untranslated region; expression of *c-fos* is very rapidly stimulated by PDGF, EGF, NGF and TPA; this is mediated through the serum response element (SRE) upstream from the start site of transcription; other members of the *fos* family include *fosB*, *Fra-1* and *Fra-2* (35–39)

Table 1. Continued

Proto-oncogene	Viral or tumor origin	Protein product	Cellular location of product	Function of protein	Human chromosome location	Tissue expression	Other information
fps/fes	Five isolates of avian sarcoma virus (FuSV-ASV, PC11-ASV, PC1V-ASV, UR1-ASV and 16L-ASV) and three isolates of feline sarcoma virus (ST-FeSV, GA-FeSV and HZ1-FeSV)	*gag-fps* 105–170 kDa proteins; *gag-fes* 85–110 kDa proteins; *c-fps/fes* (chicken) 98 kDa protein; *c-fps/fes* (mouse and human) 92 kDa protein	Cytoplasm/membrane; *v-gag-fps* may be partially associated with the cytoskeleton	Non-receptor protein-tyrosine kinase	15q25–26	Myeloid cells	*c-fps/fes* is not phosphorylated on tyrosine when isolated from the cell; activation may be in multiple independent ways, including mutation and viral transduction; there is evidence of elevated levels of expression in human tumors; *c-fps/fes* related sequences have been identified in *Drosophila* (40–42)
fyn	Identified by cross-hybridization with *src*-related sequences	59 kDa protein	Plasma membrane	Non-receptor protein-tyrosine kinase	6q21	Expressed in most tissues; high in brain and T lymphocytes	Activated by mutation and rearrangement; overexpression in fibroblasts can lead to partial transformation; *c-fyn* is implicated in proliferation control in T lymphocytes and fibroblasts (43–45)

Table 1. Continued

Proto-oncogene	Viral or tumor origin	Protein product	Cellular location of product	Function of protein	Human chromosome location	Tissue expression	Other information
gip-1	Ovary tumors and tumors of the adrenal cortex	α-subunit of G_i, the regulatory inhibitor of adenyl cyclase	Cytoplasmic face of plasma membrane	Mutation-activated form of G_i α	—	Most cells	Activated by mutation; overexpression induces the transformed phenotype in fibroblasts (46–48)
gsp	Some pituitary tumors and thyroid tumors	α-subunit of G_s, the regulatory stimulator of adenyl cyclase	Cytoplasmic face of plasma membrane	Mutation-activated form of G_s α	20	Most cells	Activated by mutation (46, 49, 50)
hst	Human stomach tumor	205 amino acid protein	—	FGF-related growth factor	11q13	—	Overexpression of cDNA transforms fibroblasts in culture; transfection with Kaposi's sarcoma DNA reveals an essentially identical oncogene; with int-2, a member of the fibroblast growth factor family (51, 52)
int-1/*wnt-1*	First identified as a chromosomal integration site for mouse mammary tumor virus in mammary tumors	Several glycoproteins, molecular weight range 36–44 kDa	Secreted and then associated with cell surface	—	12q12–13	Neural tissue in embryo; embryonal carcinoma cells; adult testes	Included in this gene family are *wnt-3* and *wnt-2*/*irp. wingless* the *Drosophila* homologue of *int-1*/*wnt-1*, a segment polarity gene; *int-1*/*wnt-1* and *int-2* cooperate during mammary carcinogenesis in doubly transgenic mice (53–56)

Table 1. Continued

Proto-oncogene	Viral or tumor origin	Protein product	Cellular location of product	Function of protein	Human chromosome location	Tissue expression	Other information
int-2	First identified as a chromosomal integration site for mouse mammary tumor virus in mammary tumors	Several proteins of molecular weight 27–32 kDa	Cell surface/extracellular matrix; modified form locates to the nucleus	Growth factor	11q13	Most embryonic tissues; embryonal carcinoma cells	*int-2* belongs to the family of fibroblast growth factors (57–60)
jun	S17 avian sarcoma virus	*gag-jun* 65 kDa protein *c-jun* 39 kDa phosphoprotein	Nucleus	Forms homo-dimers or hetero-dimers with *fos* to interact with DNA at AP1 sites; regulation of transcription	1p31–32	Expressed at low levels in most cells; rapidly induced after growth simulation by various agents	Members of a family of genes that includes *junB* and *junD*; *jun* activity may be regulated by phosphorylation; forms homo- and heterodimers via a leucine zipper motif near its C-terminus; deregulated expression of *c-jun* can transform fibroblasts; *jun* expression seems to be required for progression through the cell cycle (39, 61–63)

Table 1. Continued

Proto-oncogene	Viral or tumor origin	Protein product	Cellular location of product	Function of protein	Human chromosome location	Tissue expression	Other information
kit	Hardy Zuckerman 4 feline sarcoma virus	*gag-kit* 80 kDa protein *c-kit* 145–165 kDa glycoprotein	Plasma membrane	Truncated receptor protein-tyrosine kinase	4q11–21	Mast cells, cells of the erythroid lineage, melanocytes, germ cells and brain	*kit* codes for a receptor of the platelet-derived growth factor family. *kit* is allelic with the mouse *white spotting* locus (*W*); mutations at the *W* locus affect the proliferation, differentiation and migration of germ cells and pigment cells during development (64–66)
lck	Thymoma	56 kDa protein	Plasma membrane	Non-receptor protein-tyrosine kinase	1p35–32	Lymphoid cells	*lck* gene product interacts with CD4 and CD8 and its protein-tyrosine kinase activity is stimulated by antibody binding; homology to *src* gene (67, 68)
mas	Identified from human epidermoid carcinoma using a tumorigenicity assay	—	Plasma membrane	Angiotensin receptor	—	—	Activated by gene rearrangement and also by inappropriate expression (69, 70)

Table 1. Continued

Proto-oncogene	Viral or tumor origin	Protein product	Cellular location of product	Function of protein	Human chromosome location	Tissue expression	Other information
met	Transfection of DNA from an MNNG-transformed human osteosarcoma cell line (MNNG-HOS)	190 kDa protein	Membrane	Receptor-like protein-tyrosine kinase	7q31–32	Many human tumor cell lines	Activation involves chromosomal rearrangement; ligand binding to *met* gene product as yet unknown (71–73)
mil	Avian myelo-cytomatosis and carcinoma virus MH2	*gag-mil-myc* 100 kDa protein	Cytoplasmic	Protein-serine/threonine kinase	3p25	Many tissues	Very closely related to the *raf* gene; activation by truncation of the 5′ coding region (74)
mos	Moloney murine sarcoma virus	*env-mos* 37 kDa protein *c-mos* 40 kDa protein	Cytoplasm	Protein-serine/threonine kinase	8q11	Male and female germ cells; very low levels in most adult tissues	High levels of expression of *c-mos* transform fibroblasts efficiently but oncogenic activation may require ectopic expression; *c-mos* expression is required for meiosis in mouse and *Xenopus* oocytes; *Xenopus mos* protein phosphorylates cyclin B2 *in vitro* (75–77)

Table 1. Continued

Proto-oncogene	Viral or tumor origin	Protein product	Cellular location of product	Function of protein	Human chromosome location	Tissue expression	Other information
myb	Avian myeloblastosis (strains BAI/A and E26)	*v-myb* (AMV BAI/A) 45 kDa protein *gag-ets-myb* (AMV E26) 135 kDa protein *c-myb* 75 kDa protein phosphoprotein	Nucleus	Sequence-specific DNA-binding protein; transcriptional activator	6q22–24	Hematopoietic cells; highest levels found in early T- and B-lymphoid cells and in myeloid and erythroid cells	Activation involves deletions at both 5′ and 3′ ends of the coding region; abnormalities at the *c-myb* locus have been observed in several leukemias, colon carcinomas and melanomas. *myb* expression is cell-cycle dependent, increasing late G1 and early S; evolutionarily conserved; closely related gene found in *Drosophila* and *Zea mays* (78–81)
myc	Avian myelocytomatosis virus (strains MC29, MH2, CM11 and OK10)	*gag-myc* (MC29) 110 kDa phosphoprotein *gag-mil-myc* (MH2) 100 kDa phosphoprotein	Nucleus	DNA-binding helix-loop-helix protein; transcriptional regulator?	8q24	Different levels of expression in many different tissues	Activation involves chromosomal translocation, gene amplification, proviral insertion or retroviral transduction; part of a gene

Table 1. Continued

Proto-oncogene	Viral or tumor origin	Protein product	Cellular location of product	Function of protein	Human chromosome location	Tissue expression	Other information
		myc (MH2) 58 kDa phosphoprotein; *gag-myc* (CM11) 90 kDa phosphoprotein; *gag-pol-myc* (OK10) 200 kDa phosphoprotein; *c-myc* 62 kDa phosphoprotein					family that includes N- *myc*, L- *myc*, B- *myc* and S- *myc*; *c-myc* expression increases rapidly as quiescent cells are stimulated to re-enter the cell cycle; expression also changes during development and differentiation; both mRNA and protein of the *myc* family genes are unstable; *myc* complements activated *ras* genes to transform non-established cells fully (82–87)
neu/*erbB2* (also known as HER-2)	Rat neuro/glioblastoma	185 kDa glycoprotein	Plasma membrane	Receptor-like tyrosine-protein kinase with unknown ligand	17q21–22	Expressed in a variety of cell types but not in hemato-poietic tissues	Human and rat *neu*/*erbB2* are 50% identical at the amino acid level and essentially colinear with the EGF receptor; *neu*/*erbB2*

Table 1. Continued

Proto-oncogene	Viral or tumor origin	Protein product	Cellular location of product	Function of protein	Human chromosome location	Tissue expression	Other information
							gene is altered and amplified in many human tumors; *neu/erbB2* overexpression in breast cancer may be used as a prognostic factor; activation leading to fibroblast transformation is by mutation in the transmembrane region (88–91)
pim-1	Integration site for Moloney murine leukemia virus in T-cell lymphoma in mouse	34 kDa and 41 kDa proteins	Cytoplasmic	Protein-serine/threonine kinase	6p21	Hematopoietic tissues	Activation may involve overexpression of the normal *pim*-protein; proviral insertion provides an enhancer but also eliminated an mRNA instability sequence; Coexpression with *myc* leads to increased frequency of a tumor in transgenic mice (92, 93)

Table 1. Continued

Proto-oncogene	Viral or tumor origin	Protein product	Cellular location of product	Function of protein	Human chromosome location	Tissue expression	Other information
raf	3611 murine sarcoma virus	*gag-raf* 75 kDa protein *gag-raf* 90 kDa glycoprotein *c-raf* 75 kDa phosphoprotein	Cytoplasm; may translocate into the nucleus following mitogen treatment	Protein-serine/threonine kinase	3p25	Expressed, albeit at different levels, in all tissues and cells examined	Oncogenic activation is by amino-terminal truncation, removing a negatively regulating region; *c-raf* kinase is rapidly activated by mitogen treatment of fibroblasts, T cells and epithelial cells; *c-raf* kinase is activated by phosphorylation on serine and tyrosine; this suggests that *c-raf* may be a downstream link in the plasma membrane to nucleus signal transduction pathway; *raf* gene family includes A-*raf* and B- *raf* (94–96)

Table 1. Continued

Proto-oncogene	Viral or tumor origin	Protein product	Cellular location of product	Function of protein	Human chromosome location	Tissue expression	Other information
ras Ha-*ras*-1, Ki-*ras*-2, N-*ras*	Harvey-murine sarcoma virus (Ha-MSV); Kirsten-sarcoma virus (Ki-MSV); N-*ras* isolated from neuro-blastoma by transfection	21 kDa protein; polyisoprenyl-ated on cysteine residue close to carboxy-terminus	Inner surface of plasma membrane	Guanine nucleotide-binding proteins with GTPase activity	Ha-*ras*-1 11p15.5 Ki-*ras*-2 12p12-pter N-*ras* 1p22 and/or 1p13	Expressed in most tissues	Activation by point mutations that render them con-stitutively active by keeping the proteins in the GTP-bound state. The yeast, *Saccharomyces cerevisiae* contains two *ras* genes, *RAS1* and *RAS2*, involved in the regulation of adenylate cyclase; this does not seem to be the case in mammalian cells; GTPase-activating protein (GAP) binds *ras*p21 and is an effector of *ras* action; point mutations in *ras* family genes are detected in human carcinomas, leukemias and melanomas (97–100)

Table 1. Continued

Proto-oncogene	Viral or tumor origin	Protein product	Cellular location of product	Function of protein	Human chromosome location	Tissue expression	Other information
rel	Avian reticuloendotheliosis virus strain T	*v-rel* 59 kDa phosphoprotein *c-rel* 68 kDa protein	*v-rel* is found mostly in the cytoplasm with a fraction in the nucleus; *c-rel* is localized exclusively in the cytoplasm	DNA-binding protein; transcriptional regulator acting in a dominant negative way?	2p13–cen	High levels of expression in lymphoid cells	Activated by viral transduction; *rel* gene shares about 45% amino acid identity with the *Drosophila* gene, *dorsal*, which is a maternal effect gene that is involved in the determination of the dorsal–ventral axis in the developing embryo; found in a complex with at least three other proteins, including NF-κB (101–104)
ros	UR2 avian sarcoma virus	*gag-ros* 68 kDa phosphoprotein *c-ros* 280 kDa glycoprotein	Plasma membrane	Receptor-like protein-tyrosine kinase	6q16–22	Kidney, lung, testis; glioblastoma cell lines	*c-ros* is related to the insulin receptor gene; also related to the photoreceptor homeotic gene *sevenless* in *Drosophila*; *c-ros* (or *mcf-3*) detected using gene transfer experiments; ligand for *c-ros* not yet known (105–107)

Table 1. Continued

Proto-oncogene	Viral or tumor origin	Protein product	Cellular location of product	Function of protein	Human chromosome location	Tissue expression	Other information
sis	Simian sarcoma virus	*env-sis* 28 kDa protein *c-sis* 28 kDa protein	Membranes; secreted	Encodes the B-chain of PDGF	22q11	Smooth muscle cells, endothelial cells, mega-karyotes; also expressed in many fibro-sarcoma and glioblastoma cell lines	Activation by retroviral transduction; overexpression under the control of a retroviral promoter-induced transformation in fibroblasts (108–110)
ski	Sloan Kettering virus (avian retrovirus)	*gag-ski-pol* 110 kDa protein *gag-ski* 125 kDa protein *c-ski* 60–90 kDa proteins	Nucleus	Transcription factor?	1q12–qter	Expressed at low levels in all cell and tissue types examined	Activated by virus transduction; experiments in transgenic mice suggest *ski* may be involved in muscle differentiation; *sno* is a *ski*-related gene (111–114)
src	Rous sarcoma virus (avian)	*v-src* 60 kDa phosphoprotein *c-sr* 60 kDa phosphoprotein	Inner face of plasma membrane; targeted by myristylation at amino terminal	Non-receptor protein-tyrosine kinase	20q12–13	Neural tissues contain elevated levels; monocytes and mega-karyocytes also express high levels	C-terminal tyrosine residue important in the control of kinase activity; *c-src* forms a complex with polyoma middle T antigen; Ptdlns 3-kinase associates with *c-src* (115–118)

Table 1. Continued

Proto-oncogene	Viral or tumor origin	Protein product	Cellular location of product	Function of protein	Human chromosome location	Tissue expression	Other information
trk	Transfection of DNA from human colon carcinoma	Oncogene; 70 kDa phosphoprotein proto-oncogene; 150 kDa glycoprotein	Cytoplasm and plasma membrane	Truncated receptor-like protein-tyrosine kinase	1q31–43	Specific set of sensory neural cells	Gene rearrangement between a non-muscle tropomyosin and tyrosine kinase sequence; a second closely related gene, *trk-B*, arises from recombination with a ribosomal protein gene (119–121)
yes	Esh sarcoma virus and avian sarcoma virus Y73	*gag-yes* (ESV) 80 kDa protein *gag-yes* (Y73) 90 kDa protein *c-yes* 62 kDa protein	Inner face of plasma membrane	Non-receptor protein-tyrosine kinase	18q21.3	Expressed in many tissues but high levels found in kidney, liver, brain and retina	Member of *src* family of protein tyrosine kinases that also includes *src*, *fgr*, *fyn*, and *lck*; capable of binding to polyoma middle T antigen and activated PDGF receptors (118, 122–124)

Abbreviations: EB, Epstein–Barr; EGF, epidermal growth factor; GTP, guanosine triphosphate; NGF, nerve growth factor; PDGF, platelet-derived growth factor; TPA, 12-*O*-tetradecanoylphorbol-13-acetate; PtdIns, phosphatidylinositol.

1. SPECIFIC REFERENCES

1. Rosenberg, N. and Witte, O.N. (1988) *Adv. Virus Res.*, **35**, 39.

2. Hermans, A., Heisterkamp, N., von Lindern, M., van Baal, S., Meijer, D., van der Plas, D., Wiedeman, L.M., Groffen, J., Bootsma, D. and Grosveld, G. (1987) *Cell*, **51**, 33.

3. Cleary, M.L., Smith, C.D. and Sklar, J. (1986) *Cell*, **47**, 19.

4. Reed, C., Tsujimoto, Y., Alpers, J.D., Croce, C.M. and Nowell, P.C. (1987) *Science*, **236**, 1295.

5. Chen-Levy, Z., Nourse, J. and Cleary, M.L. (1989) *Mol. Cell. Biol.*, **9**, 701.

6. Langdon, W.Y., Hyland, C.D., Grumont, R.J. and Morse III, H.C. (1989) *J. Virol.*, **63**, 5420.

7. Miyoshi, J., Higashi, T., Mukai, H., Ohuchi, T. and Kakuhaga, T. (1991) *Mol. Cell. Biol.*, **11**, 4088.

8. Mayer, B.J., Hamaguchi, M. and Hanafusa, H. (1988) *Nature*, **332**, 272.

9. Mayer, B.J. and Hanafusa, H. (1990) *Proc. Nat. Acad. Sci. USA* , **87**, 2638.

10. Matsuda, M., Mayer B.J., Fukui, Y. and Hanafusa, H. (1990) *Science*, **248**, 1537.

11. Ron, D., Tronick, S.R., Aaronson, S.A. and Eva, A. (1988) *EMBO J.*, **7**, 2465.

12. Gratziani, G., Ron, D., Eva, A. and Srivastava, S.K. (1989) *Oncogene*, **4**, 823.

13. Sap, J., Munoz, A., Damm, K., Goldberg, Y., Ghysdael, J., Leutz, A., Beug, H. and Vennstrom, B. (1988) *Nature*, **324**, 635.

14. Weinberger, C., Thompson, C.C., Ong, E.S., Lebo, R., Gruol, D.J. and Evans, R.M. (1986) *Nature*, **324**, 641.

15. Downward, J., Yarden, Y., Mayes, E., Scrace, G., Totty, N., Stockwell, P., Ullrich, A., Schlessinger, J. and Waterfield, M.D. (1984) *Nature*, **307**, 521.

16. Lin, C.R., Chen, W.S., Kruiger, W., Stolarsky, L.S., Weber, W., Evans, R.M., Verma, I.M., Gill, G.N. and Rosenfeld, M.G. (1984) *Science*, **224**, 843.

17. Ullrich, A., Coussens, L., Hayflick, J.S., Dull, T.J., Gray, A., Tam, A.W., Lee, Y., Yarden, Y., Liebermann, T.A., Schlessinger, J., Downward, J., Mayes, E.L.V., Whittle, N., Waterfield, M.D. and Seeburg, P.H. (1984) *Nature*, **309**, 418.

18. Fujiwara, S., Fisher, R.J., Seth, A., Bhat, N.K., Showalter, S.D., Zweig, M. and Papas, T.S. (1988) *Oncogene*, **2**, 99.

19. Watson, D.K., McWilliams, M.J., Lapis, P., Lautenberger, J.A., Schweinfest, C.W. and Papas, T.S. (1988) *Proc. Natl. Acad. Sci. USA* , **85**, 7862.

20. Pognonec, P., Boulukos, K.E. and Ghysdael, J. (1989) *Oncogene*, **4**, 691.

21. Seth, A., Watson, D.K., Blair, D.G. and Papas, T.S. (1989) *Proc. Natl. Acad. Sci. USA* , **86**, 7833.

22. Mucenski, M.L., Taylor, B.A., Ihle, J.N., Hartley, J.W., Morse III, H.C., Jenkins, N.A. and Copeland, N.G. (1988) *Mol. Cell. Biol.*, **8**, 301.

23. Morishita, K., Parganas, E., Parham, D.M., Matsugi, T. and Ihle, J.N. (1990) *Oncogene*, **5**, 1419.

24. Matsugi, T., Morishita, K. and Ihle, J.N. (1990) *Mol. Cell. Biol.*, **10**, 1259.

25. Perkins, A.S., Fishel, R., Jenkins, N.A. and Copeland, N.G. (1991) *Mol. Cell. Biol.*, **11**, 2665.

26. Buchberg, A.M., Bedingian, H.G., Jenkins, N.A. and Copeland, N.G. (1990) *Mol. Cell. Biol.*, **10**, 4658.

27. O'Connell, P., Viskochil, D., Buchberg, A.M., Fountain, J., Cawthon, R.M., Culver, M., Stevens, J., Rich, D.C., Ledbetter, D.H., Wallace, M. *et al.* (1990) *Genomics*, **7**, 547.

28. Buchberg, A.M., Cleveland, L.S., Jenkins, N.A. and Copeland, N.G. (1990) *Nature*, **347**, 291.

29. Naharro, G., Robbins, K.C. and Reddy, E.P. (1984) *Science*, **223**, 63.

30. Tronick, S.R., Popescu, N.C., Cheah, M.S.C., Swan, D.C., Amsbaugh, S.C., Lengel, C.R., DiPaolo, J.A. and Robbins, K.C. (1985) *Proc. Natl. Acad. Sci. USA* , **82**, 6595.

31. Ley, T.J., Connolly, N.L., Katamine, S., Cheah, M.S.C., Senior, R.M. and Robbins, K.C. (1989) *Mol. Cell. Biol.*, **9**, 92.

32. Sherr, C.J., Rettenmier, C.W., Sacca, R. Roussel, M.F., Look, A.T. and Stanley, E.R. (1985) *Cell*, **41**, 665.

33. Sacca, R., Stanley, E.R., Sherr, C.J. and Rettenmier, C.W. (1986) *Proc. Natl. Acad. Sci. USA* , **83**, 3331.

34. Gisselbrecht, S., Fichelson, S., Sola, D., Bordereaux, D., Hampe, A., Andre, C., Galibert, F. and Tambourin, P. (1987) *Nature*, **329**, 259.

35. Greenberg, M.E. and Ziff, E.B. (1984) *Nature*, **311**, 433.

36. Treisman, P.E. (1986) *Cell*, **46**, 567.

37. Kouzarides, T. and Ziff, E.B. (1989) *Cancer Cells*, **1**, 71.

38. Woodgett, J.R. (1990) *Seminars in Cancer Biology*, **1**, 389.

39. Angel, P. and Karin, M. (1991) *Biochim. Biophys. Acta*, **1072**, 129.

40. Feldman, R.A., Hanfusa, T. and Hanafusa, H. (1980) *Cell*, **22**, 757.

41. Foster, D.A., Shibuya, M. and Hanafusa, H. (1985) *Cell*, **42**, 105.

42. Roebroek, A.J.M., Schalken, J.A., Verbeek, J.S., van den Ouweland, A.M.W., Onnekink, C., Bloemers, H.P. J. and van den Ven, W.J.M. (1985) *EMBO J.*, **4**, 2897.

43. Kawakami, T., Kawakami, Y., Aaronson, S.A. and Robbins, K.C. (1988) *Proc. Nat. Acad. Sci. USA* , **85**, 3870.

44. Semba, K., Kawai, S., Matsuzawa, Y., Yamanashi, Y., Nishizawa, M. and Toyoshima, K. (1990) *Mol. Cell. Biol.*, **10**, 3095.

45. Samelson, L.E., Phillips, A.F., Luong, E.T. and Klausner, R.D. (1990) *Proc. Natl. Acad. Sci. USA*, **87**, 4358.

46. Lyons, J., Landis, C.A., Harsh, G., Vallar, L., Grunewald, K., Feichtinger, H., Duh, Q.Y., Clark, O.H., Kawasaki, E., Bourne, H.R. and McCormick, F. (1990) *Science*, **249**, 655.

47. Lowndes, J.M., Gupta, S.K., Osawa, S. and Johnson, G.C. (1991) *J. Biol. Chem.*, **266**, 14193.

48. Gupta, S.K., Gallego, C., Lowndes, J.M., Pieiman, C.H., Sable, C., Eisfelder, B.J. and Johnson, G.L. (1992) *Mol. Cell. Biol.*, **12**, 190.

49. Landis, C.A., Masters, S.B., Spada, A., Pace, A.M., Bourne, H.R. and Vallar, L. (1989) *Nature*, **340**, 692.

50. Clementi, E., Malgaretti, N., Meldolesi, J. and Taramelli, R. (1990) *Oncogene*, **5**, 1059.

51. Sakamoto, H., Mori, M., Taira, M., Yoshida, T., Matsukawa, S., Shimizu, K., Sekiguchi, M., Terada, M. and Sugimura, T. (1986) *Proc. Natl. Acad. Sci. USA*, **83**, 3997.

52. Peters, G., Brookes, S., Smith, R., Placzek, M. and Dickson C. (1989) *Proc. Natl. Acad. Sci. USA*, **86**, 5678.

53. Nusse, R. and Varmus, H.E. (1982) *Cell*, **31**, 99.

54. Van't Veer, L.J., van Kessel, A., van Heerikhuizen, H., van Ooyen, A. and Nusse, R. (1984) *Mol. Cell. Biol.*, **4**, 2532.

55. Papkoff, J. and Schryver, B. (1990) *Mol. Cell. Biol.*, **10**, 2723.

56. Kwan, H., Pecenka, V., Tsukamoto, A., Parslow, T.G., Guzman, R., Lin, T.-P., Muller, N.J., Lee, F.S., Leder, P. and Varmus, H.E. (1992) *Mol. Cell. Biol.*, **12**, 147.

57. Dickson *et al.* (1984) *Cell*, **37**, 529.

58. Moore, R., Casey, G., Brookes, S., Dixon, S., Peters, G. and Dickson, C. (1986) *EMBO J.*, **5**, 919.

59. Ackland, P., Dixon, M., Peters, G. and Dickson, C. (1990) *Nature*, **343**, 662.

60. Muller, W.G., Lee, F.S., Dickson, C., Peters, G., Pattengale, P. and Leder, P. (1990) *EMBO J.*, **9**, 907.

61. Maki, Y., Bos, T.J., Davis, C., Starbuck, M. and Vogt, P.K. (1987) *Proc. Natl. Acad. Sci. USA*, **84**, 2848.

62. Vogt, P.K. and Bos, T.J. (1990) *Adv. Cancer Res.*, **55**, 2.

63. Kovary, K. and Bravo, R. (1991) *Mol. Cell. Biol.*, **11**, 4466.

64. Bessmer, P., Murphy, J.E., George, P.C., Qiu, F., Bergold, P.J., Lederman, L., Snyder, L.W., Brodeur, D., Zuckerman, E.E. and Hardy, W.D. (1986) *Nature*, **320**, 415.

65. Chabot, B., Stephenson, D.A., Chapman, V.M., Besmer, P. and Bernstein, A. (1988) *Nature*, **335**, 88.

66. Nocka, K., Majumder, S., Chabot, B., Ray, P., Cervone, H., Bernstein, A. and Besmer, P. (1989) *Genes and Develop.*, **3**, 816.

67. Veillette, A., Bookman, M.A., Horak, E.M. and Bolen, J.B. (1988) *Cell*, **55**, 301.

68. Veillette, A., Bookman, M.A., Horak, E.M., Samelson, L.E. and Bolen, J.B. (1989) *Nature*, **338**, 257.

69. Young, D., Waitches, G., Birchmeier, C., Fasano, O. and Wigler, M. (1986) *Cell*, **45**, 711.

70. Hunter, T. (1991) *Cell*, **64**, 249.

71. Cooper, C.S., Blair, D.G., Oskarsson, M.K., Tainsky, M.D., Eader, L.A. and Vande Woude, G.F. (1984) *Cancer Res.*, **44**, 1.

72. Dean, M., Park, M., LeBeau, M.M., Robins, T.S., Diaz, M.O., Rowley, J.D., Blair, D.G. and Vande Woude, G.F. (1985) *Nature*, **318**, 385.

73. Park, M., Dean, M., Cooper, C.S., Schmidt, M., O'Brien, S.J., Blair, D.G. and Vande Woude, G.F. (1986) *Cell*, **45**, 895.

74. Koenen, M., Sippel, A.E., Trachmann, C. and Bister, K. (1988) *Oncogene*, **2**, 179.

75. Blair, D.G., Oskarsson, M.K., Setn, A., Dunn, K.J., Dean, M., Zweig, M., Tainsky, M.A. and Vande Woude, G.F. (1986) *Cell*, **46**, 785.

76. Sagata, N., Daan, I., Oskarsson, M., Showalter, S.D. and Vande Woude, G.F. (1989) *Science*, **245**, 643.

77. Roy, L.M., Singh, B., Gautier, J., Arlinghaus, R.B., Nordeen, S.K. and Maller, J.G. (1990) *Cell*, **61**, 825.

78. Thompson, C.B., Challoner, P.B., Neiman, P.E. and Groudine, M. (1986) *Nature*, **319**, 374.

79. Biedenkapp, H., Borgmeyer, U., Sippel, A.E. and Klempnauer, K.H. (1988) *Nature*, **335**, 835.

80. Weston, K. and Bishop, J.M. (1989) *Cell*, **58**, 85.

81. Weston, K. (1990) *Seminars in Cell Biology*, **1**, 371.

82. Zimmerman, K.A., Yancopoulos, G.D., Collum, R.G., Smith, R.K., Kohl, N.E., Denis, K.A., Nau, M.M., Witte, O.N., Toran-Allerand, D., Gee, C.E., Minna, J.D. and Alt, F.W. (1986) *Nature*, **319**, 780.

83. Leder, A., Pattengale, P.K., Kuo, A., Steward, T.A. and Leder, P. (1986) *Cell*, **45**, 485.

84. Thompson, T.C., Southgate, J., Kitchener, G. and Land, H. (1988) *Cell*, **56**, 917.

85. Penn, L.J.Z., Laufer, E.M. and Laud, H. (1990) *Seminars in Cancer Biology*, **1**, 69.

86. DePinho, R.A., Schreiber-Agus, N. and Alt, F.W. (1991) *Adv. Cancer Res.*, **57**, 1.

87. Dang, C.V. (1991) *Biochim. Biophys. Acta*, **1072**, 103.

88. Bargmann, C.I., Hung, M.C. and Weinberg, R.A. (1986) *Nature*, **319**, 226.

89. Slamon, D.J., Godolphin, W., Jones, L.A., Holt, J.A., Wong, S.G., Keith, D.E., Levin, W.J., Stuart, S.G., Stuart, W.J., Udove, J., Ullrich, A. and Press, M.F. (1989) *Science*, **244**, 707.

90. Lupu, R., Colomer, R., Zugmaier, G., Sarup, J., Shepard, M., Slamon, D. and Lippman, M.E. (1990) *Science*, **249**, 1552.

91. King, C.R., Kraus, M.H., DiFiore, P.P., Paik, S. and Kasprzyk, P.G. (1990) *Seminars in Cancer Biology*, **1**, 329–337.

92. Selten, G., Cuypers, H.T., Boelens, W., Robanus-Maandag, E., Verbeek, J., Van Beveren, C. and Berns, A. (1986) *Cell*, **46**, 603.

93. van Lohuizen, J., Verbeek, S., Krimpenfort, P., Domen, J., Saris, C., Radaszkiewicz, T. and Berns, A. (1989) *Cell*, **56**, 673.

94. Schultz, A.M., Copeland, T., Orvszlan and Rapp, U.R. (1988) *Oncogene*, **2**, 187.

95. Rapp, U.R., Heidecker, G., Huleihel, M., Cleveland, J.C., Choi, W.C., Pawson, T., Ihle, J.N. and Anderson, W.B. (1989) *Cold Spring Harbor Symp. Quant. Biol.*, **53**, 173.

96. Li, P., Wood, K., Mamon, H., Haser, W. and Roberts, T. (1991) *Cell*, **64**, 479.

97. Bos, T.J. (1989) *Cancer Res.*, **49**, 4682.

98. Hall, A. (1990) *Science*, **249**, 635.

99. Marshall, C.J. (1991) *Trends in Genetics*, **7**, 91.

100. Haubruck, H. and McCormick, F. (1991) *Biochim. Biophys. Acta*, **1072**, 215.

101. Hannink, M. and Temin, H. (1989) *Mol. Cell. Biol.*, **9**, 4232.

102. Enrietto, P.T. (1990) *Seminars in Cancer Biology*, **1**, 399.

103. Ballard, D.W., Walker, W.H., Doerre, S., Sista, P., Molitor, J.A., Dixon, E.P., Peffer, N.J., Hannink, M. and Greene, W.C. (1990) *Cell*, **63**, 803.

104. Baeuerle, P.A. (1991) *Biochim. Biophys. Acta*, **1072**, 63.

105. Nekameyer, W.S. and Wang, L.-H. (1985) *J. Virol.*, **53**, 879.

106. Matsushime, H. and Shibuya, M. (1990) *J. Virol.*, **64**, 2117.

107. Birchmeier, C., O'Neil, K., Riggs, M. and Wigler, M. (1990) *Proc. Natl. Acad. Sci. USA*, **87**, 4799.

108. Josephs, S.F., Ratner, L., Clarke, M.F., Westin, E.H., Reitz, E.H. and Wong-Staal, F. (1984) *Science*, **225**, 636.

109. Gazit, A., Igarashi, H., Chiu, I.M., Srinivasan, A., Yaniv, A., Tronick, S.R., Robbins, K.C. and Aaaronson, S.A. (1984) *Cell*, **39**, 89.

110. Robbins, K.C., Leal, F., Pierce, J.H. and Aaeronson, S.A. (1985) *EMBO J.*, **4**, 1783.

111. Li, Y., Magarian, C., Teumer, J.K. and Starnezer, E. (1986) *J. Virol.*, **57**, 1065.

112. Colmenares, C. and Stavnezer, E. (1989) *Cell*, **59**, 293.

113. Sutrave, P., Kelly, A.M. and Hughes, S.H. (1990) *Genes and Develop.*, **4**, 1462.

114. Nomura, W., Sasamoto, S., Ishii, S., Date, T., Matsui, M. and Ishizaki, R. (1989) *Nucleic Acids Res.*, **17**, 5489.

115. Wyke, J.A. and Stoker, A.W. (1987) *Biochim. Biophys. Acta*, **907**, 47.

116. Hanafusa, H. (1987) *Ann. Rev. Cell Biol.*, **3**, 31.

117. Cantley, L.C., Auger, K.R., Carpenter, C., Duckworth, B., Graziani, A., Kapeller, R. and Soltoff, F. (1991) *Cell*, **64**, 281.

118. Bolen, J.B., Thompson, P.A., Eiseman, E. and Horak, I.D. (1991) *Adv. Cancer Res.*, **57**, 103.

119. Martin-Zanca, D., Oskam, R., Mitra, G., Copeland, T. and Barbacid, M. (1989) *Mol. Cell. Biol.*, **9**, 24.

120. Ziemiecki, A., Muller, R.G., Xiao-Chang, F., Hynes, N.E. and Kozma, S. (1990) *EMBO J.*, **9**, 191.

121. Martin-Zanca, D., Barbacid, M. and Parada, L.F. (1990) *Genes and Develop.*, **4**, 683.

122. Kitamura, N., Kitamura, A., Toyoshima, K., Hirayama, Y. and Yoshida, M. (1982) *Nature*, **297**, 205.

123. Sudol, M. and Hanafusa, H. (1986) *Mol. Cell. Biol.*, **6**, 2839.

124. Yamanashi, Y., Fukushige, S.I., Semba, K., Sukegawa, J., Nobuyaki, M., Matsubara, K.I., Yamamoto, T. and Toyoshima, K. (1987) *Mol. Cell. Biol.*, **7**, 237.

2. GENERAL REFERENCES

1. Carney, D. and Sikora, K. eds (1990) *Genes and Cancer.* J. Wiley, Chichester.

2. Franks, L.M. and Teich, N.M. eds (1991) *Introduction to the Cellular and Molecular Biology of Cancer (2nd edn).* Oxford University Press, Oxford.

3. Glover, D.M. and Hames, B.D. eds (1989) *Oncogenes.* IRL Press at Oxford University Press, Oxford.

4. Reddy, E.P., Skalka, A.M. and Curran, T. eds (1988) *The Oncogene Handbook.* Elsevier, Amsterdam.

5. Reviews on Oncogenic Processes (9 separate reviews) (1991) *Cell*, **64**, 235.

6. Weinberg, R.A. (1989) *Oncogenes and the Molecular Origins of Cancer.* Cold Spring Harbor Laboratory Press, New York.

CHAPTER 16
KEY DATA IN CELL BIOLOGY: CENTRIFUGATION

D. Rickwood

Centrifugation

1. CALCULATION OF CENTRIFUGAL FORCE

The centrifugal force RCF experienced by a particle depends on its radial distance from the center of rotation (r) in centimeters and the speed of rotation in r.p.m., this is true for all types of rotors.

$$\text{RCF} = 11.18 \times r \times \left(\frac{\text{r.p.m.}}{1000}\right)^2.$$

2. APPLICATIONS OF CENTRIFUGE ROTORS

Table 1. Applications of centrifuge rotors

Type of centrifuge rotor	Type of separation		
	Pelleting	**Rate-zonal (size)**	**Isopycnic (density)**
Fixed-angle	Excellent	Poor	Good
Swinging bucket	Inefficient	Excellent	Acceptable
Vertical tube	Do not use	Good	Excellent
Zonal	Do not use	Excellent	Acceptable

3. CALCULATION OF *K*-FACTORS OF ROTORS

The k-factor of a rotor is a measure of its pelleting efficiency. The k-factor is the time in hours required to pellet a particle with a sedimentation coefficient of one. It can be calculated from the minimum radius (r_{min}) and maximum radius (r_{max}) of the rotor, together with the speed of rotation (Q).

$$k = \frac{(\ln r_{max} - \ln r_{min})}{Q^2} \times 2.533 \times 10^{11}.$$

Note that as the k-factor becomes smaller the pelleting efficiency increases, as a general rule the rotor with the smallest k-factor will give you the fastest (but not always the best) separation.

4. CALCULATION OF PELLETING TIME

The time (T) in hours taken for a particles with a sedimentation coefficient of (S) to pellet in a medium with the same viscosity and density as water can be calculated from the k-factor of the rotor from the equation:

$$T=\frac{k}{S}.$$

5. DERATING ROTORS FOR USE WITH DENSE SOLUTIONS

Rotors can only be used at their maximum speed if the sample liquid does not exceed the design density of the rotor, usually 1.2 g cm^{-3}. However, some rotors are designed for higher density samples (1.7 g cm^{-3} or even 4 g cm^{-3}) and other manufacturers define the maximum load in each pocket. Derating is particularly important when using swinging-bucket rotors.

Table 2. Properties of manufacturing materials of centrifuge tubes

Material	Resistant to	Attacked by	Effective temp. range
Polycarbonate	Inorganic acids, mineral oil, vegetable oil, salts, silicone, water	Alkalis, DMSO, esters, aromatic hydrocarbons, ketones and alcohols, diethylpyrocarbonate	−10°C to +135°C
Polyallomer and polypropylene	As polycarbonate plus bases, DMSO, esters, organic acids, alcohols, glycols	Phenol	−10°C to +125°C +140°C for short periods
Corex (glass)	Most chemicals		−10°C to +140°C
Pyrex	Most chemicals		−10°C to +140°C
Polythene (rigid)	Acids, alkalies, salts, water	Organic acids, halogens, oxidizers, aromatic aliphatic and chlorinated hydrocarbons	105°C for short periods, 95°C continuous; brittle at −10°C
Nylon	Organic acids, hydrocarbons, ketones and alcohol	Oxidizers, mineral acids	
Stainless steel	All solvents	Conc. acids	−10°C to +140°C
PVC	Acids	Organic solvents, phenolic compounds	75°C for short periods 65°C continuous
Polystyrene	Acids, bases, alcohols, toluene, trichloroacetic acid	Chloroform, acetone, ether	75°C for short periods 60°C continuous
Noryl	Acids, bases, alochols	Diethyl-pyrocarbonate	0° to +130°C

The formula for calculating the amount of derating for a solution with a density of E g cm^{-3} is:

$$Q_d = Q_n \times \sqrt{\left(\frac{D}{E}\right)}$$

where Q_n is the normal maximum speed, Q_d is the maximum speed of the rotor with the dense solution and D is the design density of the rotor.

Centrifugation

Autoclavable	Flexibility	Appearance	Max. g-force	Notes
Yes	Rigid tends to crack	Transparent	300 000*g*	May go cloudy on repeated autoclaving; this does not affect the physical strength
Yes	Some	Translucent	600 000*g*	Some sources say unsuitable for below 5°C
Yes	Glass hard	Perfectly clear	30 000*g*	Stronger than Pyrex
Yes	Glass hard	Perfectly clear	5 000*g*	
No	Some	Milky white	100 000*g*	Cheap and good for low temperatures
Yes	Some	Translucent	—	Absorbs water
Yes	Metal hard	Solid	—	Rotors need derating
No	Very flexible or rigid	Translucent or opaque	—	
No	Rigid	Transparent	—	
Yes	Rigid	Opaque	200 000*g*	Insert in zonal rotors including the SZ14

6. EQUATIONS RELATING THE REFRACTIVE INDEX TO THE DENSITY OF SOLUTIONS

When solutes are dissolved in water the refractive index of the resulting solution differs from that of water. The increase in the refractive index is proportional to the concentration of solute. In the case of gradient solutes, the density of the solution is directly related to the solute concentration as well as the refractive index. This relationship can be expressed in the terms of the following equation

$$\rho = a\eta - b.$$

The tables below list the coefficients a and b for a number of ionic and non-ionic gradient media. However, before applying the equations it is essential that allowance is made for the presence of other solutes (e.g. EDTA buffers, etc.). The following equation should be used:

$$\eta_{\text{corrected}} = \eta_{\text{observed}} - (\eta_{\text{buffer}} - \eta_{\text{water}}).$$

Table 3. Ionic gradient media

Gradient solute	Temperature (°C)		Coefficients		Valid density range (g cm^{-3})
	η for	ρ	a	b	
CsCl	20	20	10.9276	13.593	1.2–1.9
	25	25	10.8601	13.497	1.3–1.9
Cs_2SO_4	25	25	12.1200	15.166	1.1–1.4
	25	25	13.6986	17.323	1.4–1.8
Cs(HCOO)	25	25	13.7363	17.429	1.7–1.8
	25	20	12.8760	16.209	1.8–2.3
NaBr	25	25	5.8880	6.852	1.0–1.5
NaI	20	20	5.3330	6.118	1.1–1.8
KBr	25	25	6.4786	7.643	1.0–1.4
KI	20	20	5.7317	6.645	1.0–1.4
	25	25	5.8356	6.786	1.1–1.7
RbCl	25	25	9.3282	11.456	1.0–1.4
RbBr	25	25	9.1750	11.241	1.1–1.7

Table 4. Non-ionic gradient media

Gradient solute	Temperature (°C)		Coefficients	
	η	ρ	a	b
Sucrose	20	0	2.7329	2.6425
Ficoll	20	20	2.381	2.175
Metrizamide	20	20	3.350	3.462
Nycodenz	20	20	3.242	3.323
Metrizoate	25	5	3.839	4.117
Renografin	24	4	3.5419	3.7198
Iothalamate	25	25	3.904	4.201
Ludox HS	20	5	8.387	10.180
Percoll	20	5	6.1905	7.252
Chloral hydrate	4	4	3.6765	3.9066
Bovine serum albumin	24	5	1.4129	0.8814

Table 5. Dilution of stock solutions of sucrose

Desired final concentration % (w/w)	Volume (ml) of stock solution to be diluted to 1 litre[a]					Density (g cm^{-3})	Refractive index of the final solution at 20°C[b]
	60% (w/w)	66% (w/w)	70% (w/v)	80% (w/v)	2.0 M		
5	66	58	73	64	75	1.0179	1.3403
10	134	119	149	130	152	1.0381	1.3479
15	205	182	228	199	233	1.0592	1.3557
20	279	247	310	271	317	1.0810	1.3639
25	357	315	396	346	405	1.1036	1.3723
30	437	387	485	424	496	1.1270	1.3811
35	521	461	578	506	591	1.1513	1.3902
40	609	539	676	591	691	1.1764	1.3997
45	701	620	777	680	795	1.2025	1.4096
50	796	704	883	773	903	1.2296	1.4200
55	896	792	994	870	—	1.2575	1.4307
60	1000	884	—	971	—	1.2865	1.4418

[a] All values quoted are at 4°C.
[b] Refractive index of water at 20°C is 1.3330. Note that the presence of solutes other than sucrose will increase the observed refractive index.

CHAPTER 17
KEY DATA IN CELL BIOLOGY: RADIOISOTOPES

D. Rickwood and D. Patel

Radioisotopes

1. DEFINITIONS

1.1. Units of radioactivity

Becquerel
The SI unit of radioactivity, 1 becquerel (Bq) is 1 distintegration per second; this is equivalent to 2.70×10^{-11} curies (Ci).

Curie (milli-, micro-)
This is equivalent to the amount of an isotope undergoing 3.7×10^{10} nuclear disintegrations per second (2.22×10^{12} per min). 1 curie equals 3.7×10^{10} becquerels.

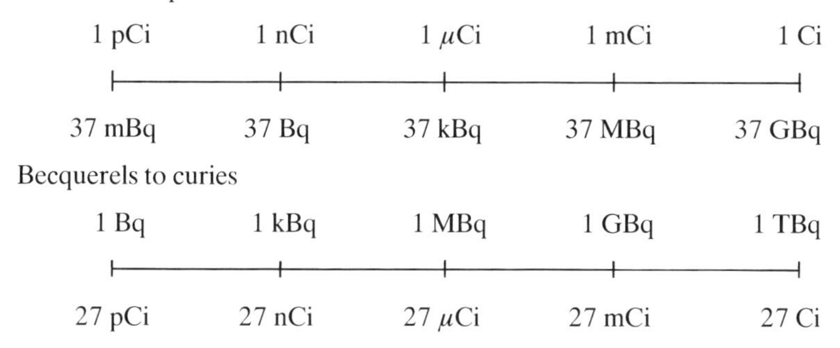

1.2. Specific activity
The radioactivity of an element per unit mass of element.

1.3. Electron volt (eV, or MeV = 10^6 eV)
Measure of energy of radioactive emissions; energy acquired by an electron when accelerating along a potential gradient of 1 volt $= 1.6 \times 10^{-12}$ erg.

1.4. Gray
This is an absorbed dose of 1 joule kg^{-1} of tissue. 1 gray equals 100 rads.

1.5. Dose equivalent man (sievert)
This is defined as: (dose in Sv) = (dose in gray) × (quality factor, Q). The value of Q is unity for β particles, X-rays and γ-rays, and 10 for α-particles, neutrons and protons. 100 rem is equal to 1 sievert (Sv).

2. RADIOACTIVE DECAY CORRECTION

$$\log_{10}(N_0/N_t) = 0.3010(t/h).$$

Percentage isotope remaining $= 100/\text{antilog}[0.3010(t/h)]$, or $100 \times e^{-0.6931(t/h)}$,

where t = time of decay;
h = half-life or time taken for radioactivity to decay to 50% of original activity;
N_0 = radioactivity at zero time;
N_t = radioactivity at time t.

After x half-lives the percentage of original isotope remaining $= 2^{-(x)} \times 100$. Thus, $(x)=1$, 50%; $(x)=2$, 25%; $(x)=3$, 12.5%; $(x)=4$, 6.25%; $(x)=5$, 3.13%; and so on.

3. AUTORADIOGRAPHY

Autoradiography has now assumed a very prominent role in both the quantitative and semi-quantitative measurement of radioisotopes in cell biology. Applications include the localization of a wide range of metabolic activities in cells to the sequencing of DNA. The method of detecting isotopically-labeled compounds depends on the isotope and the nature of the sample. The major application is the detection of radioactive materials in gels and the methods used for the commonest isotopes are given in *Table 1*.

Table 1. Sensitivities of film detection methods for commonly used radioisotopes

Isotope	Method	Detection limit d.p.m./cm^2 for 24 h	Relative performance compared to direct autoradiography
^3H	Direct autoradiography	$>8 \times 10^6$	1
(0.0186 MeV)	Fluorography using PPO	8000	>1000
^{14}C or ^{35}S	Direct autoradiography	6000	1
(0.166 MeV)	Fluorography using PPO	400	15
^{32}P	Direct autoradiography	525	1
(1.71 MeV)	Intensifying screen	50	10.5
^{125}I	Direct autoradiography	1600	1
(0.035 MeV γ rays, plus X-rays and electrons)	Intensifying screen	100	16

4. LIQUID SCINTILLATION COUNTING

4.1. Cerenkov counting

Cerenkov radiation can be used for measuring isotopes that emit a hard β-radiation (e.g. ^{32}P). No scintillator is required and, since chemical quenching is absent, the sample can simply be dissolved in a wide variety of solvent systems, such as organic solvents, perchloric acid, water, and counted directly in a scintillation counter. The efficiency is dependent on the energy of β-particles; weak β-emitters (^3H and ^{14}C) cannot be counted in this way.

Table 2. Half-lives of some radioactive isotopes

^{32}P (half-life 14.3 days)		^{35}S (half-life 87.1 days)		^{131}I (half-life 8.1 days)		^{125}I (half-life 60 days)		^{45}Ca (half-life 165 days)		^{51}Cr (half-life 28 days)	
Time (days)	% activity remaining	Time (days)	% activity remaining	Time (days)	% activity remaining	Time (days)	% activity remaining	Time (days)	% activity remaining	Time (days)	% activity remaining
1	95.3	2	98.4	0.2	98.3	4	95.5	10	95.9	2	95.2
2	90.8	5	96.1	0.4	96.6	8	91.2	20	91.9	4	90.6
3	86.5	10	92.3	0.6	95.0	12	87.1	30	88.2	6	86.2
4	82.4	15	88.7	1.0	91.8	16	83.1	40	84.5	8	82.0
5	78.5	20	85.3	1.6	87.2	20	79.4	50	81.1	10	78.1
6	74.8	25	82.0	2.3	81.2	24	75.8	60	77.7	12	74.3
7	71.2	31	78.1	3.1	76.7	28	72.4	70	74.5	14	70.7
8	67.8	37	74.5	4.0	71.0	32	69.1	80	71.5	16	67.3
9	64.7	43	71.0	5.0	65.2	36	66.0	90	68.5	18	64.0
10	61.5	50	67.0	6.1	59.3	40	63.0	100	65.7	20	61.0
11	58.7	57	63.6	7.3	53.4	44	60.2	110	63.0	22	58.0
12	55.9	65	59.6	8.1	50.0	48	57.4	120	60.4	24	55.2
13	53.2	73	56.0			52	54.8	130	57.9	26	52.5
14	50.7	81	52.5			56	52.4	140	55.5	28	50.0
14.3	50.0	87.1	50.0			60	50.0	150	53.3		
								160	51.1		

Table 3. Some radioactive isotopes used in biochemical investigations

Element	Type of radiation	Half-life	Energy of radiation (MeV)		Scintillator for detection[a]
			Particles	γ-rays	
^{3}H	β^-	12.3 yr	0.0185	—	LSC
^{14}C	β^-	5760 yr	0.156	—	LSC
^{22}Na	β^+, γ	2.6 yr	0.58(90%)	1.3	LSC or NaI(T1)
^{32}P	β^-	14.3 days	1.71	—	LSC or Cerenkov
^{35}S	β^-	87.1 days	0.169	—	LSC
^{36}Cl	β^+, K, β^-	3.1×10^5 yr	0.71	—	LSC
^{45}Ca	β^-	165 days	0.260	—	LSC
^{51}Cr	K, γ	28 days	—	0.323, 0.237	LSC or NaI(TI)
^{52}Mn	β^+(35%) K(65%)	5.8 days	0.58	1.0, 0.73 0.94, 1.46	LSC
^{54}Mn	K, γ	310 days	—	0.835	NaI(TI)
^{55}Fe	K only	2.94 yr	—	—	LSC
^{57}Co	γ	270 days	—	0.136(10%) 0.122(88%)	NaI(TI)
^{58}Co	β^+(14.5%), γ	72 days	0.472	0.805	LSC or NaI(TI)
^{59}Fe	β^-, γ	46.3 days	0.46(50%) 0.26(50%)	1.3(50%) 1.1(50%)	LSC
^{60}Co	β^-, γ	5.3 yr	0.31	1.16, 1.32	LSC or NaI(TI)
^{65}Zn	β^+(1.3%) K(98.7%)	250 days	0.32	1.14(46% of K)	LSC
^{75}Se	K, γ	121 days	—	0.076–0.405	NaI(TI)
^{86}Rb	β^-, γ	18.7 days	1.822(80%) 0.716(20%)	1.081	LSC or Cerenkov or NaI(TI)
^{90}Sr	β^-	28.5 yr	0.61	—	LSC
^{125}Sb	β^-, γ	2.7 yr	0.3(65%) 0.7(35%)	0.55	LSC
^{125}I	γ, ECcon. cl.	60 days	—	0.035	NaI(TI), LSC
^{131}I	β^-, γ	8.1 days	0.605(86%) 0.25(14%)	0.637, 0.363 0.282, 0.08	LSC or NaI(TI)

[a] LSC, liquid scintillation counting.

4.2. Counting of finely dispersed or solvent-soluble substances

Radioactive samples that are soluble in hydrocarbon solvents or finely dispersed (e.g. on filters) can be directly introduced into mixtures such as:

(i) scintillation-grade toluene or xylene containing 5 g l^{-1} 2,5-diphenyloxazole (PPO) and 0.1 g l^{-1} 1,4-di-(2-(5-phenyloxazolyl))-benzene (POPOP);
(ii) scintillation-grade toluene or xylene containing 10 g l^{-1} 2-phenyl-5-(4-biphenylyl)-1,3,4-oxadiazole (PBD) or 15 g l^{-1} 2-(4′-*t*-butylphenyl)-5-(4″-biphenylyl)-1,3,4-oxadiazole (butyl PBD).

Chemical quenching and color quenching can both occur. Particulate material will also tend to reduce the counting efficiency.

4.3. Counting of aqueous samples

Aqueous samples can be counted in hydrocarbon-based scintillation fluids using a polyethoxylate surfactant to produce an emulsion. For example:

600 ml toluene or xylene (sulfur-free or scintillation grades)
340 ml surfactant (Triton X-100 or Triton X-114)
5.0 g PPO, 0.1 g POPO — can be replaced by 15 g butyl PBD or 10 g PBD, an advantage for scintillation counters containing bialkali photocathodes.

The disadvantage of such systems is that separation into two phases can occur, with a marked reduction in counting efficiency (1). Sometimes the phase separation cannot be observed visually and it cannot be detected by external standard measurements. Consequently, each scintillation fluid must be examined for the stability of counting rate with time at the counting temperature, with various percentages of aqueous sample incorporated. A recipe for a modified Tritosol cocktail has been published (2) that can accept up to 23% of water with no phase separation and that gives a linear relationship between counting efficiency and external standard ratio over wide quenching ranges. The recipe for this is:

Ethylene glycol	35 ml
Ethanol	140 ml
Triton X-100 (X-114)	250 ml
Xylene	575 ml
PPO	3.0 g
POPOP	0.2 g

Many commercial emulsifying cocktails are also available, but it is wise to examine the counting efficiency and emulsion stability under the precise conditions being used, for example the percentage of aqueous sample incorporated, the presence of solutes, and the pH; avoid highly alkaline or acidic solutions.

Solid tissues and other solid biological samples should be solubilized before counting by incubating with quaternary bases such as (hyamine hydroxide (*p*-diisobutylcresoxyetoxyethyl)dimethylbenzylammonium hydroxide). However, this often produces strong phosphorescence with tissue proteins. Various commercial stabilizers have been developed based on quaternary bases that do not have this disadvantage. During digestion of the sample, frequent agitation or ultrasonic vibration is required: heating to 50°C helps, but higher temperatures cause excessive coloration and quenching. Alternatively, some scintillation fluids (e.g. Soluene) contain the digestion agent and can be added directly to samples.

5. METHODS OF DECONTAMINATING LABORATORIES

A survey should indicate the areas requiring decontamination and such areas should be clearly delineated; be sure to select the correct type of monitor to be used. The areas should be decontaminated promptly to prevent the spread of contamination. The decontamination procedures should be carefully planned and correct materials selected. Personnel should wear appropriate protective clothing, including plastic gloves and overshoes.

In general, wet methods should always be used to prevent the dispersal of dust. If, because of special circumstances, dry methods cannot be avoided, special precautions will be necessary (e.g. special vacuum cleaners with filters).

5.1. Methods

Spilt liquid should be absorbed on paper tissue or Vermiculite. Where dry material has been spilt and there is loose particulate contamination, the best method of decontamination may well be the application of a strippable coating by brush or spray. This coating will hold the contaminant and prevent the dispersal (provided that the method chosen does not disturb the loose contaminant). When dry, the coating is stripped off, taking the adhering contaminant with it. Self-adhesive tapes can also be applied to non-porous surfaces for the removal of loosely held dust.

Table 4. Decontamination agents

Type of surface or equipment	Decontamination agent	Treatment
Walls, floors, etc., contaminated clothing	Suitable detergents, for example Decon 90 or RBS 25, or wetting agents (it is preferable to add a little EDTA)	A first method for particularly greasy or dirty surfaces; use a 5% solution with swabbing action, or in a washing machine for clothing
Textiles, plastics, paints, rubber, metals	Detergents or combinations of citric acid, EDTA, etc.	Use in the form of a cream or immerse articles in tanks for periods up to several hours (0.8% solution at elevated temperatures)
Linoleum	Organic solvents, detergents	To remove the normal waxed coating
Machine tools	Solvents (proprietary grease or emulsifying solvents)	Apply directly to heavily oiled and greasy surfaces with a cloth or brush, emulsifying solvents may be rinsed off with water
Glassware	Detergents, concentrated acids, chromic acid	Use in the normal way
Stainless steel, mild steel and light alloys, ferrous metals	Sulfuric acid, sulfuric acid with inhibitors, nitric acid/ sodium-fluoride, proprietary rust removers	These dissolve the contaminated surface, taking it into solution
Painted surfaces	Paint removers, solvent strippers, alkaline strippers	Use in difficult cases where the paint itself has to be removed

Repeat any of the above if necessary, but if this does not remove the contamination satisfactorily, apply mild abrasive cleaners (proprietary brands) with a cloth to the affected surface. Frequently the inclusion of unlabeled compounds of the same type as the contaminant in the decontamination solution will facilitate decontamination.

For the second stage of decontamination, the use of damp swabs is preferable to uncontrolled sluicing, to prevent spread of contamination. The actual method to be used and the appropriate agent will depend to some extent on the ease of removal of the remaining contamination and the methods set out below are arranged in order, to deal with increasingly difficult circumstances:

(i) Treat with a suitable detergent, which may be in the form of a cream to prevent the spread of the contamination by splashing. Swabbing or light scrubbing may also be necessary.

(ii) Scrub lightly with a complexing solution. If an application of a thickened complexing agent is used, it should be left on the surface for a few hours before rinsing off. The addition of pigment will help to identify the areas to which the decontamination procedure has been applied.

(iii) Swab or scrub with mild abrasive pastes containing complexing agents.

(iv) Where none of the above methods is successful and the contamination still remains, it will be necessary to treat the surface by more vigorous scrubbing and abrasion or more severe treatment, for example by planing off wood surfaces and chipping away concrete and brick surfaces. In such cases it will be necessary to restore the original surface before further work is recommenced.

Radioisotopes

6. REFERENCES

1. Pande, S.V. (1976) *Anal. Biochem.*, **74**, 25.

2. Fricke, U. (1975) *Anal. Biochem.*, **63**, 555.

INDEX